Lecture Notes in Mathematics

For information about Vols. 1–1201 please contact your bookseller or Springer-Verlag

Vol. 1202: A. Dür, Möbius Functions, Incidence Algebras and Power Series Representations. XI, 134 pages. 1986.

Vol. 1203: Stochastic Processes and Their Applications. Proceedings, 1985. Edited by K. Itô and T. Hida. VI, 222 pages. 1986.

Vol. 1204: Séminaire de Probabilités XX, 1984/85. Proceedings. Edité par J. Azéma et M. Yor. V, 639 pages. 1986.

Vol. 1205: B.Z. Moroz, Analytic Arithmetic in Algebraic Number Fields. VII, 177 pages. 1986.

Vol. 1206: Probability and Analysis, Varenna (Como) 1985. Seminar. Edited by G. Letta and M. Pratelli. VIII, 280 pages. 1986.

Vol. 1207: P.H. Bérard, Spectral Geometry: Direct and Inverse Problems. With an Appendix by G. Besson. XIII, 272 pages. 1986.

Vol. 1208: S. Kaijser, J.W. Pelletier, Interpolation Functors and Duality. IV, 167 pages. 1986.

Vol. 1209: Differential Geometry, Peñíscola 1985. Proceedings. Edited by A.M. Naveira, A. Ferrández and F. Mascaró. VIII, 306 pages. 1986.

Vol. 1210: Probability Measures on Groups VIII. Proceedings, 1985. Edited by H. Heyer. X, 386 pages. 1986.

Vol. 1211: M.B. Sevryuk, Reversible Systems. V, 319 pages. 1986.

Vol. 1212: Stochastic Spatial Processes. Proceedings, 1984. Edited by P. Tautu. VIII, 311 pages. 1986.

Vol. 1213: L.G. Lewis, Jr., J.P. May, M. Steinberger, Equivariant Stable Homotopy Theory. IX, 538 pages. 1986.

Vol. 1214: Global Analysis – Studies and Applications II. Edited by Yu.G. Borisovich and Yu.E. Gliklikh. V, 275 pages. 1986.

Vol. 1215: Lectures in Probability and Statistics. Edited by G. del Pino and R. Rebolledo. V, 491 pages. 1986.

Vol. 1216: J. Kogan, Bifurcation of Extremals in Optimal Control. VIII, 106 pages. 1986.

Vol. 1217: Transformation Groups. Proceedings, 1985. Edited by S. Jackowski and K. Pawalowski. X, 396 pages. 1986.

Vol. 1218: Schrödinger Operators, Aarhus 1985. Seminar. Edited by E. Balslev. V, 222 pages. 1986.

Vol. 1219: R. Weissauer, Stabile Modulformen und Eisensteinreihen. III, 147 Seiten. 1986.

Vol. 1220: Séminaire d'Algèbre Paul Dubreil et Marie-Paule Malliavin. Proceedings, 1985. Edité par M.-P. Malliavin. IV, 200 pages. 1986.

Vol. 1221: Probability and Banach Spaces. Proceedings, 1985. Edited by J. Bastero and M. San Miguel. XI, 222 pages. 1986.

Vol. 1222: A. Katok, J.-M. Strelcyn, with the collaboration of F. Ledrappier and F. Przytycki, Invariant Manifolds, Entropy and Billiards; Smooth Maps with Singularities. VIII, 283 pages. 1986.

Vol. 1223: Differential Equations in Banach Spaces. Proceedings, 1985. Edited by A. Favini and E. Obrecht. VIII, 299 pages. 1986.

Vol. 1224: Nonlinear Diffusion Problems, Montecatini Terme 1985. Seminar. Edited by A. Fasano and M. Primicerio. VIII, 188 pages. 1986.

Vol. 1225: Inverse Problems, Montecatini Terme 1986. Seminar. Edited by G. Talenti. VIII, 204 pages. 1986.

Vol. 1226: A. Buium, Differential Function Fields and Moduli of Algebraic Varieties. IX, 146 pages. 1986.

Vol. 1227: H. Helson, The Spectral Theorem. VI, 104 pages. 1986.

Vol. 1228: Multigrid Methods II. Proceedings, 1985. Edited by W. Hackbusch and U. Trottenberg. VI, 336 pages. 1986.

Vol. 1229: O. Bratteli, Derivations, Dissipations and Group Actions on C*-algebras. IV, 277 pages. 1986.

Vol. 1230: Numerical Analysis. Proceedings, 1984. Edited by J.-P. Hennart. X, 234 pages. 1986.

Vol. 1231: E.-U. Gekeler, Drinfeld Modular Curves. XIV, 107 pages. 1986.

Vol. 1232: P.C. Schuur, Asymptotic Analysis of Soliton Problems. VIII, 180 pages. 1986.

Vol. 1233: Stability Problems for Stochastic Models. Proceedings, 1985. Edited by V.V. Kalashnikov, B. Penkov and V.M. Zolotarev. VI, 223 pages. 1986.

Vol. 1234: Combinatoire énumérative. Proceedings, 1985. Edité par G. Labelle et P. Leroux. XIV, 387 pages. 1986.

Vol. 1235: Séminaire de Théorie du Potentiel, Paris, No. 8. Directeurs: M. Brelot, G. Choquet et J. Deny. Rédacteurs: F. Hirsch et G. Mokobodzki. III, 209 pages. 1987.

Vol. 1236: Stochastic Partial Differential Equations and Applications. Proceedings, 1985. Edited by G. Da Prato and L. Tubaro. V, 257 pages. 1987.

Vol. 1237: Rational Approximation and its Applications in Mathematics and Physics. Proceedings, 1985. Edited by J. Gilewicz, M. Pindor and W. Siemaszko. XII, 350 pages. 1987.

Vol. 1238: M. Holz, K.-P. Podewski and K. Steffens, Injective Choice Functions. VI, 183 pages. 1987.

Vol. 1239: P. Vojta, Diophantine Approximations and Value Distribution Theory. X, 132 pages. 1987.

Vol. 1240: Number Theory, New York 1984–85. Seminar. Edited by D.V. Chudnovsky, G.V. Chudnovsky, H. Cohn and M.B. Nathanson. V, 324 pages. 1987.

Vol. 1241: L. Gårding, Singularities in Linear Wave Propagation. III, 125 pages. 1987.

Vol. 1242: Functional Analysis II, with Contributions by J. Hoffmann-Jørgensen et al. Edited by S. Kurepa, H. Kraljević and D. Butković. VII, 432 pages. 1987.

Vol. 1243: Non Commutative Harmonic Analysis and Lie Groups. Proceedings, 1985. Edited by J. Carmona, P. Delorme and M. Vergne. V, 309 pages. 1987.

Vol. 1244: W. Müller, Manifolds with Cusps of Rank One. XI, 158 pages. 1987.

Vol. 1245: S. Rallis, L-Functions and the Oscillator Representation. XVI, 239 pages. 1987.

Vol. 1246: Hodge Theory. Proceedings, 1985. Edited by E. Cattani, F. Guillén, A. Kaplan and F. Puerta. VII, 175 pages. 1987.

Vol. 1247: Séminaire de Probabilités XXI. Proceedings. Edité par J. Azéma, P.A. Meyer et M. Yor. IV, 579 pages. 1987.

Vol. 1248: Nonlinear Semigroups, Partial Differential Equations and Attractors. Proceedings, 1985. Edited by T.L. Gill and W.W. Zachary. IX, 185 pages. 1987.

Vol. 1249: I. van den Berg, Nonstandard Asymptotic Analysis. IX, 187 pages. 1987.

Vol. 1250: Stochastic Processes – Mathematics and Physics II. Proceedings 1985. Edited by S. Albeverio, Ph. Blanchard and L. Streit. VI, 359 pages. 1987.

Vol. 1251: Differential Geometric Methods in Mathematical Physics. Proceedings, 1985. Edited by P.L. García and A. Pérez-Rendón. VII, 300 pages. 1987.

Vol. 1252: T. Kaise, Représentations de Weil et GL_2 Algèbres de division et GL_n. VII, 203 pages. 1987.

Vol. 1253: J. Fischer, An Approach to the Selberg Trace Formula via the Selberg Zeta-Function. III, 184 pages. 1987.

Vol. 1254: S. Gelbart, I. Piatetski-Shapiro, S. Rallis. Explicit Constructions of Automorphic L-Functions. VI, 152 pages. 1987.

Vol. 1255: Differential Geometry and Differential Equations. Proceedings, 1985. Edited by C. Gu, M. Berger and R.L. Bryant. XII, 243 pages. 1987.

Vol. 1256: Pseudo-Differential Operators. Proceedings, 1986. Edited by H.O. Cordes, B. Gramsch and H. Widom. X, 479 pages. 1987.

Vol. 1257: X. Wang, On the C*-Algebras of Foliations in the Plane. V, 165 pages. 1987.

Vol. 1258: J. Weidmann, Spectral Theory of Ordinary Differential Operators. VI, 303 pages. 1987.

continued on page 115

Lecture Notes in Mathematics

Edited by A. Dold, B. Eckmann and F. Takens

Subseries: Scuola Normale Superiore, Pisa
Adviser: E. Vesentini

1424

Francis E. Burstall
John H. Rawnsley

Twistor Theory for Riemannian Symmetric Spaces

With Applications to Harmonic Maps of Riemann Surfaces

Springer-Verlag

Berlin Heidelberg New York London Paris Tokyo Hong Kong

Authors

Francis E. Burstall
School of Mathematical Sciences
University of Bath
Bath BA2 7AY, Great Britain

John H. Rawnsley
Mathematics Institute
University of Warwick
Coventry CV4 7AL, Great Britain

Mathematics Subject Classification (1980): Primary: 58E20, 53C30, 53C35, 53C55
Secondary: 83C60

ISBN 3-540-52602-1 Springer-Verlag Berlin Heidelberg New York
ISBN 0-387-52602-1 Springer-Verlag New York Berlin Heidelberg

Printing and binding: Druckhaus Beltz, Hemsbach/Bergstr.
2146/3140-543210 – Printed on acid-free paper

Cstack
SCIMON

Table of Contents

Introduction

Background

The subject of this monograph is the interaction between real and complex homogeneous geometry and its application to the study of minimal surfaces (or harmonic maps). That minimal surfaces may be studied by complex variable methods is by no means a new idea since it informs the work of Weierstrass on minimal surfaces in $\mathbf{R}^3$. However, we shall take as our starting point the seminal papers of Calabi [23, 24] in the late sixties. In those papers, Calabi investigated minimal immersions of a 2-sphere in S^{2n} by associating to each such immersion a holomorphic curve in the homogeneous Kähler manifold $\mathbf{SO}(2n+1)/\mathbf{U}(n)$. The methods of Complex Analysis were then brought to bear on these auxiliary holomorphic curves. The fruits of this analysis include a complete classification of all minimal 2-spheres in an n-sphere in terms of certain holomorphic 2-spheres in a complex projective space and a quantisation of the area of such minimal surfaces.

These ideas were taken up in the eighties by a number of physicists and mathematicians and a similar analysis of harmonic maps (i.e. branched minimal immersions) of a 2-sphere in a complex projective space was soon provided [12, 29, 34, 36]. Again, a key step is to associate to each such harmonic map a holomorphic map of S^2 into an auxiliary complex manifold, in this case a flag manifold of the form $\mathbf{U}(n+1)/\mathbf{U}(r)\times\mathbf{U}(1)\times\mathbf{U}(n-r)$. Since then there has been a great deal of activity in extending these results to other co-domains such as Grassmannians [1, 21, 22, 28, 59, 83], Lie groups [72, 73, 85] and other classical symmetric spaces [82, 3].

Meanwhile, in 1984, Eells-Salamon [33] observed that more flexibility could be obtained by considering pseudo-holomorphic curves in certain non-integrable almost complex manifolds. Indeed, they associated such a curve to any conformal harmonic map of a Riemann surface into an oriented Riemannian 4-manifold. Now pseudo-holomorphic curves are much less easy to handle (see, however, [39]) but, nonetheless, these ideas provided a useful framework in which many of the previous results could be understood.

Finally, in 1985, Uhlenbeck's analysis of harmonic 2-spheres in the unitary group $\mathbf{U}(n)$ appeared [72]. Here again, a decisive role is played by holomorphic curves in an auxiliary complex space but this time the space in question is the infinite-dimensional Kähler manifold of based loops in $\mathbf{U}(n)$: the loop group $\Omega\mathbf{U}(n)$. One of Uhlenbeck's main results is the existence of a Bäcklund transform, repeated application of which produces all harmonic 2-spheres in $\mathbf{U}(n)$ from the constant maps. This provides a classification of all such 2-spheres which subsumes and extends most of the known results in this direction for harmonic 2-spheres in complex Grassmannians.

Throughout a large part of the above development, attention has focussed on harmonic maps into certain Riemannian symmetric spaces together with holomorphic curves in associated homogeneous Kähler manifolds. This monograph has its genesis in our attempt to understand the relationship provided by these ideas between the geometry of symmetric spaces and the geometry of complex homogeneous spaces.

Before turning to a discussion of this relationship, we refer the Reader who wishes to learn more of the matters so briefly touched upon above to the surveys [16, 17, 84].

Overview

There are three main topics that are treated in this work: homogeneous geometry, twistor theory and harmonic maps. Let us describe each of these in turn:

Homogeneous Geometry

We deal with two classes of reductive homogeneous spaces: the Riemannian symmetric spaces and the (generalised) flag manifolds. The first class have been extensively studied and need no introduction here. The flag manifolds are comparatively less well-known (see, however, [5, 80]) although they have a rich geometry and exhaust the compact Kähler homogeneous spaces with semi-simple isometry group [6]. One of our main aims is to demonstrate a close relationship that exists between flag manifolds and the Riemannian symmetric spaces with inner involution (we call these *inner* symmetric spaces). To be more precise, we show that each flag manifold G/H fibres homogeneously over an inner symmetric space G/K in an essentially unique way. Moreover, each inner symmetric space is the target of such a fibration for at least one flag manifold (and generally more than one). These fibrations, which we call *canonical* fibrations, are defined entirely by the algebra of the situation: specifically, by the data of a parabolic subgroup of a complex semi-simple Lie group and a compact real form of that Lie group.

We shall also identify an invariant holomorphic distribution (the *superhorizontal* distribution) which is transverse to the fibres of our canonical fibrations. This distribution enjoys the property that holomorphic curves tangent to it project onto minimal surfaces under the canonical fibrations. In favourable circumstances, all minimal 2-spheres in the symmetric space arise in this way.

A useful tool in this development is a particular realisation of the flag manifold as an adjoint orbit (the orbit of the *canonical element*) which, although completely natural, appears to be quite different from other such realisations discussed by Borel [65].

It is perhaps surprising that the non-compact version of these ideas is rather better known. In fact, there is substantial overlap between the non-compact analogue of our theory and the theory of period matrix domains [38]. In particular, in that setting, the superhorizontal distribution is just that which defines the infinitesimal period relation.

These fibrations also appear in the thesis of W. Schmid [64] in his study of irreducible representations of non-compact semisimple Lie groups obtained on cohomology groups of holomorphic line bundles over flag domains.

Twistor theory

A central role in the twistor theory of a Riemannian manifold N is played by the bundle $J(N) \to N$ of almost Hermitian structures on N. This bundle carries a natural almost complex structure (denoted here by J_1) which is, however, rarely integrable. In a search for complex manifolds associated to N, we consider the zero set Z of the Nijenhuis tensor of J_1, which is *a priori* a set with very little structure. However, when N is an inner symmetric space, we shall show that the isometry group of N acts transitively on each connected component of Z and that each such component is in fact a flag manifold holomorphically embedded in $J(N)$. Moreover, the trace of the bundle projection on each component is a canonical fibration and all canonical fibrations are realised in this way exactly once. This gives a geometrical interpretation of the algebraic constructions discussed above and, at the same time, completely elucidates the structure of Z. It seems clear that this theory will have applications to the geometry of inner symmetric spaces. We shall present some preliminary results in this direction and refer the Reader also to

[19].

The structure theorem for Z is closely related to some results of Bryant [11] who considered the intersection of Z and the zero set of the obstruction to holomorphicity of the horizontal distribution of $J(N)$. This smaller zero set again has flag manifolds for components although of a rather restricted type (in the terminology of chapter 4, they have height not exceeding two). Thus our results may be viewed as an extension of those of Bryant. Our methods, however, are quite different and provide a new proof of Bryant's theorem.

Harmonic maps

As applications of our theory, we study harmonic maps of S^2 into a symmetric space N. Our results may be split into three categories.

Firstly, we show that if N is inner then any such harmonic map is the image under a canonical fibration of a pseudo-holomorphic curve in a flag manifold. Here the flag manifold is equipped with a non-integrable almost complex structure *à la* Eells-Salamon but we show that, under certain conditions, the vanishing of a holomorphic differential guarantees that the pseudo-holomorphic curves are in fact holomorphic for the standard complex structure on the flag manifold. This provides a uniform proof of results of Calabi [24], Eells-Wood [34] and Bryant [10] concerning harmonic 2-spheres in S^{2n}, $\mathbf{C}P^n$ and $\mathbf{H}P^1$ respectively. Much of the previous theory of minimal 2-spheres has depended on the vanishing of a series of holomorphic differentials. In our approach, we identify a universal holomorphic differential which is also of use in our study of stable harmonic 2-spheres, to which we now turn.

We completely characterise the stable harmonic 2-spheres in a simply-connected irreducible symmetric space of compact type N. It turns out that the results depend solely on $\pi_2(N)$ which is either zero, $\mathbf{Z}_2$ or $\mathbf{Z}$ (in case that N is Hermitian symmetric). If N is Hermitian symmetric, we obtain an *a priori* proof of the result that stable harmonic 2-spheres are $\pm$-holomorphic; a result originally proved by Siu-Zhong [66, 87] by checking cases. If $\pi_2(N)$ vanishes, we show that all stable harmonic 2-spheres are constant. The remaining case, when $\pi_2(N) = \mathbf{Z}_2$, is rather more interesting. here N contains a family of totally geodesically immersed Hermitian symmetric subspaces (which turn out to be projective spaces) with the property that maps factoring holomorphically through one are stable and harmonic as maps into N. Moreover, we show that any stable harmonic 2-sphere in N must so factor. It is an interesting corollary of this development that such a symmetric space must admit a null-homotopic stable harmonic 2-sphere which is non-constant. These results form joint work with Simon Salamon and we thank him for his permission to let us present them here.

Finally, we extend the results of Uhlenbeck [72] mentioned above to a large class of simple Lie groups G (those with Hermitian symmetric quotient). In so doing, we follow Valli's insightful approach [73] to Uhlenbeck's work. We find a Bäcklund transform for harmonic maps $S^2 \to G$ with which all such maps may be constructed from the constant maps. This answers to a large extent question 2 posed by Uhlenbeck in [72]. As an application of these ideas, we obtain a number of results concerning gap phenomena for harmonic 2-spheres in symmetric spaces some of which are new.

Remark on methods

The Reader may be dismayed but not surprised to learn that we have recourse to a large amount of Lie theory during this work. Perhaps more novel and interesting is our repeated use of the theorem of Birkhoff-Grothendieck on the decomposition of holomorphic vector bundles and the reduction of holomorphic principal bundles on S^2. Indeed, chapters 2 and 6–8 may be read as an essay on the applications of this theorem to the study of harmonic 2-spheres. This idea is not new: in the investigation of stable maps, its use may be traced back to Siu-Yau [67], while, in studies of the construction of harmonic maps, it makes its first fleeting appearance in the work of Erdem-Wood [35]. However, we hope to demonstrate in this work that its use as a powerful and unifying tool in this area has been underestimated. In particular, this appears to be the first time that the full force of the Grothendieck version of the theorem [40] has been applied to harmonic maps.

Table of contents

As a guide to the Reader, we present a brief description of the contents of each chapter.

Chapter 1 contains generalities about homogeneous geometry and sets up our approach to that subject. The main point of interest is that we define a Lie algebra valued 1-form on a reductive homogeneous space which satisfies an analogue of the Maurer-Cartan equations. Much of the geometry of such a space can be described using this 1-form and in this way we provide a framework for calculating on homogeneous spaces which is of great use in the sequel.

Chapter 2 introduces harmonic maps and we apply the Birkhoff-Grothendieck theorem for the first time to demonstrate the existence of pseudo-holomorphic curves in $J(N)$ covering a harmonic map $S^2 \to N$ where N is an even-dimensional manifold. Extensions of this result to suitably ramified minimal surfaces of higher genus in Kähler manifolds via the Harder-Narasimhan filtration are also presented.

Chapter 3 introduces symmetric spaces. Here also are collected various items of structure theory which we need in the sequel. The main result of the chapter is the determination of the second homotopy group of the irreducible symmetric spaces of compact type. We provide a simple root-theoretic criterion for determining this group together with an explicit set of generators. Our approach is based on Murakami's version [53] of the classification of involutions of a compact simple Lie algebra. The Reader with no taste for structure theory is invited to skip most of this.

Chapter 4 is one of the most central in the monograph. After some algebraic preliminaries, we define flag manifolds and their non-compact analogues, the flag domains. We then construct the canonical fibrations and superhorizontal distributions and give some examples. Our construction throws up a particular choice of Kähler metric on a flag manifold and we discuss the cohomology class that the Kähler form represents. This last will be needed in chapter 8. Finally, an appendix describes the adjustments to be made for non-inner symmetric spaces.

Chapter 5 is consecrated to the zero set of the Nijenhuis tensor on $J(N)$ and the determination of its structure.

Chapter 6 describes the covering of harmonic 2-spheres in symmetric spaces by pseudo-holomorphic curves in flag manifolds. Just as in chapter 2 we use the Birkhoff-Grothendieck theorem for this. We also discuss holomorphic differentials.

Chapter 7 contains our classification of stable harmonic 2-spheres in compact symmetric spaces.

Chapter 8 is concerned with our Bäcklund transform procedure for harmonic 2-spheres in a simple Lie group and its applications.

Final remarks and acknowledgments

Some of the work described in chapters 4, 6 and 8 was announced in [18] while the results of chapter 7 were announced in [20]. Readers with long memories should also note that a preliminary version of this manuscript was in circulation under the title *Twistors, Homogeneous*

Geometry and Harmonic Maps.

During the lengthy course of the preparation of this monograph, we have benefited from innumerable conversations with too many colleagues to mention. Special thanks are due to Simon Salamon, especially for his collaboration in the results of chapter 7; Georgio Valli, for informing us of his work and John C. Wood and Jim Eells for their advice and encouragement.

This document was typeset by the first author on a SUN-3 workstation running under UNIX using the typesetting program *troff* together with a local (to Bath) enhancement of the *eqn* mathematics preprocessor written by Professor R. Sibson. The first author would like to thank Professor Sibson for sharing his expertise in this area.

Chapter 1. Homogeneous Geometry

Let M be a manifold on which is given a smooth transitive action of a Lie group G. Choosing a base-point $x_0 \in M$ we let H be the stability subgroup of x_0 and then we have a principal H-bundle $\pi : G \to M$, where $\pi(g) = g.x_0$, on which G acts by left translations as bundle automorphisms. Indeed M is diffeomorphic to G/H and then π is just the coset fibration. The surjective map of the Lie algebra $\mathbf{g}$ of G, $\mathbf{g} \to T_{x_0}M$ given by

$$\xi \mapsto \left.\frac{d}{dt}\right|_{t=0} \exp t\xi . x_0$$

has the Lie algebra $\mathbf{h}$ of H as kernel and so induces an isomorphism of $\mathbf{g}/\mathbf{h}$ with $T_{x_0}M$. We extend this by equivariance to get an isomorphism of the associated bundle $G \times_H \mathbf{g}/\mathbf{h}$ with TM which is given explicitly by

$$[g, \xi + \mathbf{h}] \mapsto g_* \left.\frac{d}{dt}\right|_{t=0} \exp t\xi . x_0 = \left.\frac{d}{dt}\right|_{t=0} \exp t \mathrm{Ad} g \xi . x$$

where $x = \pi(g)$.

Notation. If W is a representation of H we shall henceforth denote the associated bundle $G \times_H W$ by $[W]$.

The homogeneous space M is said to be *reductive* if the Lie algebra $\mathbf{g}$ has a decomposition $\mathbf{g} = \mathbf{h} \oplus \mathbf{m}$ for some $\mathrm{Ad}_G H$-invariant summand $\mathbf{m}$. Then $\mathbf{g}/\mathbf{h} \cong \mathbf{m}$ as H-spaces so $[\mathbf{g}/\mathbf{h}] \cong [\mathbf{m}]$ and hence we have an isomorphism $[\mathbf{m}] \cong TM$. Since $\mathbf{m}$ is an invariant subspace of $\mathbf{g}$ we have an inclusion $[\mathbf{m}] \subset [\mathbf{g}]$. The latter may be canonically identified with the trivial bundle $\underline{\mathbf{g}} = \mathrm{M} \times \mathbf{g}$ via

$$[g, \xi] \mapsto (\pi(g), \mathrm{Ad} g \xi).$$

Thus we have an identification of TM with a subbundle of the trivial bundle which we may view as a $\mathbf{g}$-valued 1-form β on M.

If $P_{\mathbf{m}}$ (resp. $P_{\mathbf{h}}$) denotes the projection onto $\mathbf{m}$ (resp. $\mathbf{h}$) then we have

$$\beta_x(X) = \mathrm{Ad} g P_{\mathbf{m}}(\mathrm{Ad} g^{-1} \xi) \quad \text{if} \quad X = \left.\frac{d}{dt}\right|_{t=0} \exp t\xi . x \tag{1}$$

and $x = \pi(g)$. From this it is easy to see that β is equivariant in the sense that

$$g^* \beta = \mathrm{Ad} g \beta.$$

Note the useful identity

$$X = \left.\frac{d}{dt}\right|_{t=0} \exp t \beta_x(X) . x, \quad X \in T_x M.$$

Example. If M is actually the group manifold G, acting on itself by left translations, then we see from the above identity that β is just the (right) Maurer-Cartan form of G.

By analogy with the above example, we shall call β the *Maurer-Cartan form* of the reductive homogeneous space M.

A homogeneous space M of G is called a *symmetric* space if there is an involution τ_0 of G with $(G^{\tau_0})_0 \subset H \subset G^{\tau_0}$. Then $\mathbf{h}=\mathbf{g}^{\tau_0}$ and $\mathbf{m}=\mathbf{g}^{-\tau_0}$ is a reductive summand which satisfies the additional condition $[\mathbf{m},\mathbf{m}]\subset\mathbf{h}$. Symmetric spaces are the most widely studied class of reductive homogeneous spaces. We shall see the geometrical significance of the extra condition on the reductive summand in this special case shortly.

Returning to the general situation, the left translation of a reductive summand $\mathbf{m}$ around G provides a G-invariant distribution which is horizontal for π and right H-invariant. This defines a G-invariant connection in the principal bundle $\pi : G \to M$. This procedure produces a bijective correspondence between reductive summands $\mathbf{m}$ and G-invariant connections in $\pi : G \to M$. We shall view the reductive homogeneous space M as coming equipped with a fixed summand $\mathbf{m}$ and refer to the corresponding connection as the *canonical connection.* Its connection form α (as an $\mathbf{h}$-valued 1-form on G) is $P_{\mathbf{h}}\theta$ where θ is the left-invariant Maurer-Cartan form of G. G-invariant tensors on M (or, more generally, G-invariant sections of associated bundles) are parallel with respect to the canonical connection and in particular the canonical connection is a metric connection for any invariant metric on M. Given such a metric, TM becomes isomorphic with the cotangent bundle T^*M and so is a symplectic G-space. β is essentially the momentum map for this symplectic action of G.

The canonical connection induces a covariant differentiation D in any associated bundle $[V]$ for any representation V of H. If V happens to be the restriction of a representation of G then $[V]$ can be identified with the trivial bundle $\underline{V}$ via the map

$$[g,\mathbf{v}] \mapsto (\pi(g), g.\mathbf{v}).$$

In this case there is a simple relationship between flat differentiation and the covariant differentiation induced on $\underline{V}$ by the canonical connection.

Proposition 1.1. *Let $f: M \to \underline{V}$ be a smooth section of $\underline{V}$ then*

$$df = Df + \beta.f \quad .$$

Proof. Under the identification of $\underline{V}$ with $[V]$ a section f of $\underline{V}$ corresponds with an H-equivariant map $\hat{f}: G \to V$ as follows

$$\hat{f}(g) = g^{-1}.f(\pi(g)) \ . \tag{2}$$

Further, the covariant differential Df lifts as the V-valued 1-form $d\hat{f}+\alpha.\hat{f}$ on G. Differentiating (2) and using the Leibniz rule gives

$$d\hat{f}_g = -\theta.g^{-1}.f\circ\pi + g^{-1}.\pi^*df \quad .$$

Thus

$$\begin{aligned}(d\hat{f}+\alpha.\hat{f})_g &= g^{-1}.\pi^*df + (\alpha-\theta)_g.\hat{f} \\ &= g^{-1}.\pi^*df - P_{\mathbf{m}}\theta.\hat{f} \\ &= g^{-1}(\pi^*df - \mathrm{Ad}g(P_{\mathbf{m}}\theta).f\circ\pi).\end{aligned}$$

Now, pulling equation (1) back to G we have

$$(\pi^*\beta)_g = \mathrm{Ad}g(P_{\mathbf{m}}\theta)$$

whence

$$d\hat{f} + \alpha.\hat{f} = g^{-1}.\pi^*(df - \beta.f)$$

and so

$$Df = df - \beta.f$$

concluding the proof. □

Suppose W is an H-invariant subspace of the G-representation V. Then $[W]$ is a G-invariant subbundle of the trivial bundle $\underline{V}$ and hence the inclusion $[W] \subset \underline{V}$ is parallel for the canonical connection. Thus one way to view proposition (1.1) is as a formula for the canonical connection in $[W]$. If $V = W_1 + W_2$ is an H-invariant splitting of a representation of G then

$$\underline{V} = [W_1] + [W_2]$$

is a G-invariant splitting of $\underline{V}$ into D-parallel subbundles. As a simple consequence of proposition (1.1) we have the following:

Lemma 1.2 *Denote by $\sigma: G \to End(V)$ the representation of G on V and let $\underline{V} = [W_1] + [W_2]$. Let $P_i: \underline{V} \to [W_i]$ be the projection onto $[W_i]$ viewed as a function $P_i: M \to End(V)$. Then*

$$dP_i = [\sigma(\beta), P_i].$$

Proof. The projections are G-invariant and hence D-parallel. But on $\underline{End(V)}$

$$D = d - [\sigma(\beta), .],$$

whence the result. □

Example. Letting σ be the adjoint representation of G we see that

$$\underline{\mathbf{g}} = [\mathbf{h}] + [\mathbf{m}] \quad .$$

As we remarked above $[\mathbf{m}] = TM$ while we see that $[\mathbf{h}]_{\pi(g)} \subset \underline{\mathbf{g}}$ is $\mathrm{Ad}\pi(g)\mathbf{h}$ and this is the isotropy Lie algebra at $\pi(g)$. Hence we call $[\mathbf{h}]$ the *isotropy bundle*. Henceforth we will denote the projection onto the tangent bundle by $P: \underline{\mathbf{g}} \to [\mathbf{m}]$. Let us observe that for $\xi \in \mathbf{g}$, if we define a vector field $\tilde{\xi}$ by

$$\tilde{\xi}_x = \frac{d}{dt}\bigg|_{t=0} \exp t\xi . x \;, \qquad x \in M,$$

then formula (1) immediately gives us

$$\beta(\tilde{\xi}) = P\xi. \tag{3}$$

As remarked above, the 1-form β is the analogue for reductive homogeneous spaces of the Maurer-Cartan form for Lie groups. We now prove the analogue of the structure equations for β.

Lemma 1.3.

$$d\beta = (1 - \tfrac{1}{2}P)[\beta \wedge \beta].$$

Proof. We begin by differentiating the equivariance relation

$$g^*\beta = \mathrm{Ad}g\beta$$

to obtain

$$L_{\tilde{\xi}}\beta = [\xi, \beta].$$

Applying Cartan's Identity we get

$$i_{\tilde{\xi}}d\beta = [\xi, \beta] - d(i_{\tilde{\xi}}\beta).$$

From (3) we have that

$$i_{\tilde{\xi}}\beta = P\xi$$

so that we obtain from lemma (1.2)

$$d(i_{\tilde{\xi}}\beta) = d(P\xi) = (dP)\xi = [\mathrm{ad}\beta, P]\xi.$$

Thus

$$\begin{aligned} i_{\tilde{\xi}}d\beta &= [\xi, \beta] - [\beta, P\xi] + P[\beta, \xi] \\ &= (1-P)[\xi, \beta] - [\beta, P\xi] \\ &= (1-P)[P\xi, \beta] + [P\xi, \beta] \\ &= (2-P)[P\xi, \beta]. \end{aligned}$$

But

$$\begin{aligned} i_{\tilde{\xi}}[\beta \wedge \beta] &= [\beta(\tilde{\xi}), \beta] - [\beta, \beta(\tilde{\xi})] \\ &= 2[P\xi, \beta], \end{aligned}$$

whence the result follows since the $\tilde{\xi}_x$ span T_xM for all $x \in M$. □

As a simple corollary let us compute the torsion and curvature of the canonical connection:

Corollary 1.4. *Let T, R denote the torsion and curvature respectively of the canonical connection. Then*

$$\beta \circ T = -\tfrac{1}{2}P[\beta \wedge \beta]$$
$$\beta \circ R = -\tfrac{1}{2}[(1-P)[\beta \wedge \beta], \beta].$$

Proof. From proposition (1.1) we have

$$\beta(D_XY) = X\beta(Y) - [\beta(X), \beta(Y)]$$

since we are in the adjoint representation. Thus

$$\begin{aligned} \beta(T(X,Y)) &= \beta(D_XY - D_YX - [X,Y]) \\ &= d\beta(X,Y) - 2[\beta(X), \beta(Y)], \end{aligned}$$

i.e.

$$\beta \circ T = d\beta - [\beta \wedge \beta]$$

and the formula for the torsion follows immediately from lemma (1.3).

As for the curvature,

$$\begin{aligned} \beta(R(X,Y)Z) &= \beta(D_XD_YZ - D_YD_XZ - D_{[X,Y]}) \\ &= [d_X - \mathrm{ad}\beta(X), d_Y - \mathrm{ad}\beta(Y)]\beta(Z) - [X,Y]\beta(Z) + [\beta[X,Y], \beta(Z)] \end{aligned}$$

$$= -\text{ad}d\beta(X,Y)\,\beta(Z) + \text{ad}[\beta(X),\beta(Y)]\,\beta(Z),$$

whence

$$\beta\circ R = \tfrac{1}{2}\text{ad}[\beta\wedge\beta]\circ\beta - \text{ad}d\beta\circ\beta$$

and the result now follows from lemma (1.3). □

Remark. It is clear from Corollary (1.4) that the canonical connection is torsion-free if and only if $[\mathbf{m},\mathbf{m}]\subset\mathbf{h}$, that is, if and only if

$$\mathbf{g} = \mathbf{h}+\mathbf{m}$$

is a symmetric decomposition of $\mathbf{g}$. Further, if M has an invariant metric, the above condition is precisely that the canonical and Levi-Civita connections coincide. Thus these connections coincide for symmetric spaces. Returning now to the general setting of lemma (1.2), we observe that flat differentiation followed by the projection P_i defines a connection in $[W_i]$. It is natural to ask when either of these connections coincides with the canonical one.

Lemma 1.5. *Let* $\underline{V}=[W_1]+[W_2]$ *be a G-invariant splitting and let* D_1 *be the connection in* $[W_1]$ *given by*

$$D_1 = P_1\circ d.$$

Then

$$D_1 = d - P_2\sigma(\beta)\ .$$

Further, $D_1 = D$ *if and only if* $\mathbf{m}.W_1\subset W_2$ *and in this case* $\sigma(\beta)\,|\,[W_1]$ *is the second fundamental form of* $[W_1]$ *in* $\underline{V}$.

Proof. We have from lemma (1.2),

$$\begin{aligned} D_1 = P_1\circ d = d - dP_1 &= d-[\sigma(\beta),P_1] \\ &= d-\sigma(\beta)+P_1\sigma(\beta) \\ &= d-P_2\sigma(\beta). \end{aligned}$$

Thus $D_1 = D$ if and only if

$$\sigma(\beta) = P_2\sigma(\beta)$$

on $[W_1]$, if and only if

$$P_1\sigma(\beta)P_1 = 0,$$

or, by equivariance, if and only if

$$\mathbf{m}.W_1\subset W_2.$$

Lastly, we note that the second fundamental form of $[W_i]$ in $\underline{V}$ is given by

$$P_2\circ d\,|\,C^\infty([W_1]) = P_2\circ\sigma(\beta)\,|\,C^\infty([W_1])$$

and the lemma follows. □

Remark. Applying this to $\underline{\mathbf{g}}=[\mathbf{h}]+[\mathbf{m}]$ we see that the canonical connection on $TM=[\mathbf{m}]$ coincides with $P\circ d$ if and only if $[\mathbf{m},\mathbf{m}]\subset\mathbf{h}$. Thus for Riemannian symmetric spaces all three connections we have considered coincide. This will be of importance when we come to study maps into symmetric spaces, for it gives us two simple formulas for the Levi-Civita connection, as either the canonical connection, or the projection of flat differentiation in the ambient trivial bundle. An important special case is the Lie group G itself which we consider next as an example.

Example The group G may be viewed as a symmetric space for $G\times G$ with action

$$(g_1,g_2)g = g_1 g g_2^{-1}$$

so that the isotropy subgroup at e is the diagonal subgroup $H=\{(g,g): g\in G\}$ and the involution on $G\times G$ is given by

$$\tau_0(g_1,g_2) = (g_2,g_1)$$

Thus $H=(G\times G)^{\tau_0}$ and

$$\mathbf{m} = (\mathbf{g}+\mathbf{g})^{-\tau_0} = \{(\xi,-\xi): \xi\in\mathbf{g}\} \equiv \mathbf{g}$$

To calculate β, take $g\in G$, $X\in T_gG$ and then we have

$$\begin{aligned}\beta_g(X) &= \beta_g((g,e)_*(g_{-1},e)_*X)\\ &= ((g,e)^*\beta)_g(L_{g^{-1}*}X)\\ &= \mathrm{Ad}(g,e)(\beta_e(L_{g^{-1}*}X))\\ &= \mathrm{Ad}(g,e)(\tfrac{1}{2}L_{g^{-1}*}X,-\tfrac{1}{2}L_{g^{-1}*}X)\\ &= (\tfrac{1}{2}R_{g^{-1}*}X,-\tfrac{1}{2}L_{g^{-1}*}X).\end{aligned}$$

Thus if θ^L, θ^R are the right- and left-invariant Maurer-Cartan forms,

$$\beta = (\tfrac{1}{2}\theta^R,-\tfrac{1}{2}\theta^L).$$

Substituting this formula into that of proposition (1.1) and projecting onto the second factor gives us

$$\theta^L(D_XY) = X(\theta^L(Y)) + \tfrac{1}{2}[\theta^L(X),\theta^L(Y)]$$

so that identifying TG with $\underline{\mathbf{g}}$ via θ^L we see that

$$D = d + \tfrac{1}{2}\mathrm{ad}\theta^L.$$

Thus our canonical connection is the $\frac{1}{2}$-connection of Cartan-Schouten.

Returning to general reductive homogeneous spaces M, if $\varphi:N\to M$ is a smooth map of a manifold N into our homogeneous space, it is easy to see that our constructions are functorial. Indeed, the (straightforward) proof of the following lemma is left to the reader.

Lemma 1.6. *Let $\sigma:G\to\mathrm{End}(V)$ be a representation, $\underline{V}$ the trivial bundle and $\varphi:N\to M$ a smooth map. Then $\varphi^{-1}\underline{V}$ is trivial and the pull-back of the canonical connection on $\underline{V}$ (over N) is given by*

$$\varphi^{-1}D = d - \sigma(\varphi^*\beta).$$

As an application of our methods let us now prove the well known result that a symmetric space G/K may be totally geodesically immersed in G (c.f. [27]). More generally, we compute the condition for the differential of an equivariant map φ of reductive homogeneous spaces to be parallel for the canonical connection.

Proposition 1.7. *Let N_i, $i=1,2$, be reductive homogeneous G_i-spaces with reductive summands $\mathbf{m}_i$ and $\varphi: N_1 \to N_2$ be a map equivariant with respect to a homomorphism $\rho: G_1 \to G_2$ so*

$$\varphi(gx) = \rho(g)\varphi(x)$$

for $x \in N_1$, $g \in G_1$. Then $d\varphi$ is parallel with respect to the canonical connections if and only if, at some $x \in N_1$,

$$[(1-\varphi^{-1}P_2)\rho([\mathbf{m}_1]_x), \varphi^{-1}P_2\rho([\mathbf{m}_1]_x)] \equiv 0.$$

Here $P_2: \underline{\mathbf{g}} \to [\mathbf{m}_2]$ is projection along the isotropy bundle of N_2.

Remark. This condition is satisfied if $\rho([\mathbf{m}_1]_x) \subset [\mathbf{m}_2]_{\varphi(x)}$.

We begin the proof with a lemma of independent interest:

Lemma 1.8. *Under the hypotheses of (1.7), let β_i denote the Maurer-Cartan form of N_i. Then*

$$\varphi^*\beta_2 = \varphi^{-1}P_2 \circ \rho \circ \beta_1 .$$

Proof. Let $X \in T_x N_1$ then

$$X = \frac{d}{dt}\bigg|_{t=0} \exp t\beta_1(X).x,$$

so

$$\begin{aligned} d\varphi(X) &= \frac{d}{dt}\bigg|_{t=0} \varphi(\exp t\beta_1(X).x) \\ &= \frac{d}{dt}\bigg|_{t=0} \rho(\exp t\beta_1(X)).\varphi(x) \\ &= \frac{d}{dt}\bigg|_{t=0} \exp t\rho(\beta_1(X)).\varphi(x) \end{aligned}$$

whence

$$\varphi^*\beta_2(X) = (\varphi^{-1}P_2)\rho\beta_1(X).$$

□

Proof of (1.7). If D^1, D^2 are the canonical connections we must show that

$$Dd\varphi(X,Y) = \varphi^{-1}D^2_X d\varphi(Y) - d\varphi(D^1_X Y) \equiv 0.$$

Using (1.8), we have

$$\begin{aligned} \varphi^{-1}\beta_2(Dd\varphi(X,Y)) &= \varphi^{-1}D^2_X(\varphi^{-1}P_2\rho\beta_1(Y)) - \varphi^{-1}P_2\rho\beta_1(D^1_X Y) \\ &= \varphi^{-1}P_2\{\varphi^{-1}D^2_X(\rho\beta_1(Y)) - \rho\beta_1(D^1_X Y)\} \end{aligned}$$

since $D^2 P_2 \equiv 0$, so

$$\begin{aligned} \varphi^{-1}\beta_2(Dd\varphi(X,Y)) &= -\varphi^{-1}P_2\{[\varphi^*\beta_2(X), \rho\beta_1(Y)] - \rho[\beta_1(X), \beta_1(Y)]\} \\ &= -\varphi^{-1}P_2[(\varphi^{-1}P_2 - 1)\rho\beta_1(X), \rho\beta_1(Y)] \\ &= [(1-\varphi^{-1}P_2)\rho\beta_1(X), \varphi^{-1}P_2\rho\beta_1(Y)]. \end{aligned}$$

The proposition now follows by evaluating the above formula at any point $x \in N_1$, which is sufficient by equivariance. □

Corollary 1.9. *Let G/K be a symmetric space with involution τ_0 at eK. Then we have a totally geodesic immersion $\varphi: G/K \to G$ defined by*

$$\varphi(gK) = g^{\tau_0} g^{-1}.$$

Proof. Define $\rho: G \to G\times G$ by $\rho(g) = (g^{\tau_0}, g)$. Then φ is equivariant with respect to ρ and for $\xi \in \mathbf{m}$

$$\rho(\xi) = (\tau_0 \xi, \xi) = (-\xi, \xi)$$

whence ρ preserves the symmetric decompositions of G/K and $G = G\times G/G$ and so φ is totally geodesic by lemma (1.7). □

Let us conclude this chapter with an example to illustrate the concepts we have introduced:

Let $G_{r,n}$ denote the complex Grassmannian of r-planes in $\mathbf{C}^n$. Clearly, the unitary group $\mathbf{U}(n)$ acts transitively on $G_{r,n}$ so that $G_{r,n}$ is a homogeneous space. Let us take as a base point V_0, the span of the first r elements of the canonical basis of $\mathbf{C}^n$. Then the stabiliser H of V_0 is isomorphic to $\mathbf{U}(r)\times\mathbf{U}(n-r)$. Further, letting τ_0 denote the involutive automorphism of $\mathbf{U}(n)$ obtained by conjugating with

$$J_r = \begin{cases} 1 & \text{on } V_0 \\ -1 & \text{on } V_0^{\perp} \end{cases}$$

we see that $H = \mathbf{U}(n)^{\tau_0}$ so that $G_{r,n}$ is a symmetric space. The Lie algebra $\mathbf{u}(n)$ of $\mathbf{U}(n)$ is the algebra of $n\times n$ skew-Hermitian matrices and the corresponding symmetric decomposition is given by

$$\mathbf{u}(n) = \mathbf{h} \oplus \mathbf{m}$$

where

$$\mathbf{h} = \{A \in \mathbf{u}(n):\ AV_0 \subset V_0\}$$
$$\mathbf{m} = \{A \in \mathbf{u}(n):\ AV_0 \subset V_0^{\perp},\ \ AV_0^{\perp} \subset V_0\}$$

Now, we have

$$[V_0]_{gV_0} = gV_0 \quad \text{for } g \in \mathbf{U}(n),$$

so that $[V_0] \subset \underline{\mathbf{C}}^n$ is the tautological bundle $T \to G_{r,n}$ whose fibre at $W \in G_{r,n}$ is W itself. Similarly,

$$[\mathbf{m}]_W = \{A \in \mathbf{u}(n):\ AW \subset W^{\perp},\ \ AW^{\perp} \subset W\}.$$

Thus, the trivial bundle $\underline{\mathbf{C}}^n$ admits a $\mathbf{U}(n)$-invariant splitting as the direct sum of the tautological bundle and its orthocomplement. Moreover, by lemma (1.5), the Maurer-Cartan form of $G_{r,n}$ is the sum of the second fundamental forms of T and $T^{\perp}$. This enables us to reduce many questions concerning the geometry of $G_{r,n}$ to the study of sub-bundles of $\underline{\mathbf{C}}^n$ and their second fundamental forms: an approach which recently proved to be useful in the study of harmonic maps into Grassmannians (c.f. [22, 21]).

Finally, an easy computation shows that the totally geodesic embedding of $G_{r,n}$ into $\mathbf{U}(n)$ provided by lemma (1.9) is just $J_r(1-2P)$. So, after a left translation by J_r, we see that the family of Grassmannians $G_{r,n}$, $r=0,\ldots,n$ are realised totally geodesically in $\mathbf{U}(n)$ as the components of the set

$$\{g\in\mathbf{U}(n):\ g^2=1\}$$

as was observed by Uhlenbeck [72].

Chapter 2. Harmonic maps and twistor spaces

Let (M,g) and (N,h) be Riemannian manifolds and $\varphi: M \to N$ a smooth map. The *energy* of φ, denoted $E(\varphi)$, is given by

$$E(\varphi) = \tfrac{1}{2}\int_M \mathrm{trace}_g\, \varphi^* h\, dvol_M .$$

A map is *harmonic* if it extremizes the energy with respect to all compactly supported variations. The associated Euler-Lagrange equation is

$$\tau_\varphi \equiv \mathrm{trace}_g \nabla d\varphi = 0$$

where ∇ is the connection on $T^*M \otimes \varphi^{-1}TN$ induced by the Levi-Civita connections of M and N. The quantity $\tau_\varphi \in C^\infty(\varphi^{-1}TN)$ is called the *tension field.*

For surveys of harmonic map theory see [31, 32].

In the case the domain is two-dimensional, there are a number of special features to the theory: the energy is conformally invariant for the domain metric, and the tension field has a particularly simple form. Indeed, if $z=x+iy$ is a local complex coordinate on M then the tension field is given, up to a conformal factor, by

$$(\varphi^{-1}\nabla^N)_{\frac{\partial}{\partial \bar z}}\, \varphi_*(\frac{\partial}{\partial z}).$$

This is in fact a Cauchy-Riemann equation as the following theorem shows:

Theorem 2.1. [48] *Let $E \to M$ be a complex vector bundle over a Riemann surface M with connection ∇. Then there is a unique holomorphic structure on E compatible with ∇, that is: a local section σ of E is holomorphic if and only if $\nabla_{\bar Z}\sigma = 0$ for all (1,0) vectors Z.*

We call this the *Koszul-Malgrange* holomorphic structure induced by ∇.

Writing $d\varphi = dz \otimes \delta + d\bar z \otimes \bar\delta$ we obtain the well-known

Proposition 2.2. *Let M be a Riemann surface and $\varphi: M \to N$ a map. Then φ is harmonic if and only if $dz \otimes \delta$ is a holomorphic section of $\kappa_M \otimes \varphi^{-1}TN^{\mathbf{C}}$.*

Here κ_M denotes the canonical line bundle of M. Since non-trivial holomorphic sections have only isolated zeros we can make the

Definition. If $\varphi: M \to N$ is a harmonic map of a Riemann surface, the *ramification index* of φ, denoted by r_φ, is the number of zeros of $dz \otimes \delta$ counted with their multiplicities.

Remark. If φ is not constant, r_φ is finite and non-negative. In this case we denote by L_φ the line subbundle of $\varphi^{-1}TN^{\mathbf{C}}$ spanned locally by δ where the latter is non-zero. It follows that $dz \otimes \delta$ is a non-zero section of the line bundle $\kappa_M \otimes L_\varphi$ and hence that $r_\varphi = c_1(\kappa_M \otimes L_\varphi)[M]$.

Following work of Calabi [23], Eells-Wood [34] and more recently Eells-Salamon [33] and Rawnsley [61], it is natural to attempt to study harmonic maps of surfaces by relating them to holomorphic maps (which may have values in an associated almost complex manifold). With this in mind we make the following definition:

Definition. A twistor fibration $\pi:Z\to N$ (with twistor space Z) is a fibration of an almost complex manifold Z over a Riemannian manifold N with the following property:

If M is almost Hermitian with co-closed Kähler form and $\psi:M\to Z$ is holomorphic then $\pi\circ\psi:M\to N$ is harmonic.

Remark. The above definition of a twistor space is chosen purely for its relevance to harmonic map theory and differs from others in the literature (for instance we do not demand that the fibres be complex submanifolds of Z).

Example [33]. Let N be a $2n$-dimensional Riemannian manifold and let $\pi:J(N)\to N$ be the bundle of Hermitian almost complex structures on N. Thus

$$J_x(N) = \{J\in \mathrm{End}(T_xN): J^2=-1,\ J \text{ skew-symmetric}\}.$$

This bundle is associated to the orthonormal frame bundle of N with typical fibre $O(2n)/U(n)$ which is a Hermitian symmetric space. Thus the vertical distribution $V=\ker\pi_*$ inherits an almost complex structure J^V. Further the Levi-Civita connection on N induces a splitting

$$TJ(N) = V\oplus H$$

where $H\cong\pi^{-1}TN$ and thus acquires a tautological almost complex structure J^H given by

$$J_j^H = j.$$

This gives two almost complex structures on $J(N)$: $J_1=J^V+J^H$ and $J_2=(-J^V)+J^H$. This second almost complex structure J_2 makes $\pi:J(N)\to N$ into a twistor fibration. To see this, first observe that a map $\psi:M\to J(N)$ with $\pi\circ\psi=\varphi$ is the same as an almost complex structure on $\varphi^{-1}TN$ or, equivalently, a maximally isotropic subbundle $\underline{\psi}$ (the (1,0) vectors) of $\varphi^{-1}TN^{\mathbb{C}}$. We have

Proposition 2.3. [61] *Let $\psi:M\to J(N)$ be a map of an almost Hermitian manifold and set $\varphi=\pi\circ\psi$. Then ψ is holomorphic with respect to J_1 if and only if*

$(i)_1$ $\varphi^{-1}\nabla^N_{\bar{Z}}C^\infty(\underline{\psi})\subset C^\infty(\underline{\psi})$, $Z\in C^\infty(T^{1,0}M)$;

(ii) $\varphi_*(T^{1,0}M)\subset\underline{\psi}$,

and holomorphic with respect to J_2 if and only if

$(i)_2$ $\varphi^{-1}\nabla^N_{Z}C^\infty(\underline{\psi})\subset C^\infty(\underline{\psi})$, $Z\in C^\infty(T^{1,0}M)$;

(ii) $\varphi_*(T^{1,0}M)\subset\underline{\psi}$.

Remarks. (a) Conditions (i) and (ii) correspond with the holomorphicity of the vertical and horizontal parts of the differential of ψ respectively.

(b) If $\dim M=2$ we see from (2.1) that $(i)_2$ is equivalent to $\underline{\psi}$ being a holomorphic subbundle of $\varphi^{-1}TN^{\mathbb{C}}$, while (ii) is equivalent to $L_\varphi\subset\underline{\psi}$.

Now we have

Theorem 2.4. [33] $\pi:(J(N),J_2)\to N$ *is a twistor fibration.*

Proof. Let M be an almost Hermitian manifold with co-closed Kähler form and let $Z_1,\ldots,Z_m$ be a local orthonormal frame field for $T^{1,0}M$. Then the tension field of φ is given by

$$\tau_\varphi = \sum \nabla d\varphi(\bar{Z}_i,Z_i) = \sum \varphi^{-1}\nabla^N_{\bar{Z}_i}\varphi_* Z_i - \sum \varphi_* \nabla^M_{\bar{Z}_i} Z_i .$$

The condition on the Kähler form ensures that $\sum \nabla^M_{\bar{Z}_i} Z_i \in T^{1,0}M$ so that $(i)_2$ and (ii) of proposition (2.3) imply that both summands of τ_φ are contained in $\underline{\psi}$. Thus τ_φ is both real and isotropic and hence zero. □

Clearly it is desirable to know which harmonic maps are projections of J_2-holomorphic maps. For this we restrict attention to Riemann surfaces as domains. Since $\varphi^{-1}TN$ has an almost complex structure coming from such a ψ, it is necessarily an oriented bundle and so $w_1(\varphi^{-1}TN)=0$. This gives a topological restriction on φ, namely $\varphi^* w_1(N)=0$. There is a geometrical restriction also, for (ii) of proposition (2.3) implies that φ is (weakly) conformal. In fact these are the only restrictions, a result which is implicit in theorem 9.10 of [61]. For completeness, we give a proof here:

Theorem 2.5. *A map* $\varphi:M\to N^{2n}$ *of a Riemann surface has a* J_2*-holomorphic lift* $\psi:M\to J(N)$ *if and only if it is weakly conformal, harmonic and* $\varphi^* w_1(N)=0$.

Proof. We have already seen that projections of J_2-holomorphic maps are weakly conformal harmonic with $\varphi^* w_1(N)=0$.

For the converse, let φ be weakly conformal and harmonic and L_φ the holomorphic line bundle locally spanned by δ. Since φ is weakly conformal, L_φ is isotropic. We must extend L_φ to a maximally isotropic holomorphic sub-bundle $\underline{\psi}$ of $\varphi^{-1}TN^{\mathbf{C}}$ and then the corresponding map $\psi:M\to J(N)$ will be our desired lift by (2.3).

To extend an isotropic holomorphic sub-bundle V of rank k to one of rank $k+1$ amounts to finding an isotropic holomorphic line sub-bundle of $E=V^\circ/V$, where V° is the polar of V. For this, it suffices to find an isotropic meromorphic section of E. Such a section is a solution of a quadratic equation in rank E unknowns over the function field of meromorphic functions on M. According to Lang [49], such a solution exists if rank $E\geq 3$, i.e. if $2n-2k\geq 3$. Thus, starting with L_φ, we can find successive extensions until we have a rank $(n-1)$ isotropic holomorphic bundle V containing L_φ.

Now V°/V has rank 2 and will admit an isotropic line sub-bundle if and only if it is orientable. This, in turn, happens precisely when $\varphi^{-1}TN$ is orientable which is equivalent to the vanishing of $\varphi^* w_1(N)=0$. Lastly, it is easy to check that a line sub-bundle of a rank 2 holomorphic bundle that is isotropic with respect to a holomorphic non-degenerate bilinear form is itself holomorphic. The theorem now follows. □

This result is an abstract existence theorem and does not give a means to construct a J_2-holomorphic lift explicitly. In order to find a more practical means for constructing lifts (which we do by finding holomorphic maximal isotropic subbundles of $\varphi^{-1}TN^{\mathbf{C}}$) we have recourse to the Harder-Narasimhan filtration of a holomorphic vector bundle on a Riemann surface which we now describe.

Definition. Let $E \to M$ be a complex vector bundle. The *slope* of E is defined as the quotient $\mu(E) = \deg E/\mathrm{rank}\, E$. Here $\deg E = c_1(E)[M]$. If E is holomorphic it is said to be *semi-stable* if for all non-zero holomorphic subbundles F of E we have $\mu(F) \leq \mu(E)$. E is *stable* if strict inequality holds for proper subbundles.

A good supply of semi-stable bundles is guaranteed by:

Theorem 2.6. [41] *Let $E \to M$ be a holomorphic vector bundle over a Riemann surface, then there exists a filtration of E by holomorphic subbundles*

$$0 = E_0 \subset E_1 \subset \ldots \subset E_r = E$$

with each quotient $D_i = E_i/E_{i-1}$ semi-stable and $\mu(D_1) > \mu(D_2) > \ldots > \mu(D_r)$.

The subbundles E_i are uniquely determined by these conditions, and we call this the *Harder-Narasimhan (HN) filtration* of E.

Following Atiyah-Bott [2], (which the reader should consult for more details) we make the following definitions

a) $\sup E = \mu(D_1)$, the maximum slope of a semi-stable quotient in the filtration;

b) $\inf E = \mu(D_r)$, the minimum slope of the semi-stable quotients.

E is semi-stable if and only if $\sup E = \inf E$, while $\sup E - \inf E$ is a measure of the instability of E.

We collect some results on the HN filtration which form the basis for its usefulness in our constructions.

Lemma 2.7. [2] *Let E, F be holomorphic vector bundles over a Riemann surface. Then*

$$\inf(E \otimes F) = \inf E + \inf F;$$

and, further, $\inf E \geq \mu$ *if and only if for all F with* $\sup F < \mu$ *there are no nonzero holomorphic bundle morphisms $E \to F$.*

Notation. Henceforth it will be convenient to write the HN filtration in a slightly different way by repeating factors if necessary so that

$$\ldots \subset E_1 \subset E_0 \subset \ldots \subset E$$

and E_i/E_{i+1} is zero or else semistable of slope i. With these conventions we have $\inf E_i \geq i$ and $\sup E/E_i < i$.

Being intrinsically defined, the HN filtration is compatible with any extra structure which E carries. For example:

Proposition 2.8. (c.f. [2]) *Let $E \to M$ be a holomorphic vector bundle with a nondegenerate holomorphic symmetric bilinear form on its fibres, then the polar E_i° of E_i in the HN filtration is E_{1-i}. In particular E_1 is isotropic.*

Proof. It is easy to see that E is semistable if and only if E^* is and that

$$\ldots \subset E_i^\perp \subset E_{i+1}^\perp \subset \ldots \subset E^*$$

has all the properties of the HN filtration of E^* where $E_i^{\perp}$ is the set of all elements of E^* which vanish on E_i. But the inner product gives an isomorphism of E with E^* which carries $E_i{}^\circ$ to $E_i^{\perp}$. By uniqueness of the HN filtration it follows $E_i{}^\circ = E_{1-i}$. □

We shall see another argument of this type below when we consider holomorphic bundles of Lie algebras.

Now let $\varphi: M \to (N,h)$ be a smooth map of a Riemann surface into a Kähler manifold. Put $T = \varphi^{-1}TN^{\mathbf{C}}$ and $D = \varphi^{-1}\nabla^N$, the pull-back of the Levi-Civita connection on N. Since N is Kähler we have the D-parallel decomposition of T into types:

$$T = T' \oplus T''.$$

Giving T the Koszul-Malgrange holomorphic structure induced by D we have the HN filtration of T

$$\ldots T_1 \subset T_0 \subset \ldots \subset T$$

and compatible filtrations of T' and T'' so that

$$T_i = T'_i \oplus T''_i.$$

Further the induced complex metric on T from h is D-parallel and so holomorphic, whence

$$T_i{}^\circ = T_{1-i}$$

by lemma (2.8).

Lemma 2.9. *For each i, $T'_i \oplus T''_{1-i}$ is a maximally isotropic subbundle of T.*

Proof. $T'_i = T' \cap T_i$, so $T'_i{}^\circ = T' + T_{1-i}$ and $T''_{1-i}{}^\circ = T'' + T_i$, whence

$$(T'_i \oplus T''_{1-i})^\circ = (T' + T_{1-i}) \cap (T_i + T'') = T'_i \oplus T''_{1-i}$$

which gives the result. □

In particular, $T'_0 \oplus T''_1 = \underline{\psi}$ and $T'_1 \oplus T''_0 = \underline{\psi}'$ define maps $\psi', \psi'': M \to J(N)$ which satisfy the vertical part of the J_2-holomorphic condition. To get the horizontal part of the condition we have to see that δ lies in $\underline{\psi}$ or $\underline{\psi}'$. This will certainly be the case if φ is sufficiently ramified:

Theorem 2.10. *Let M_g be a closed Riemann surface of genus g and $\varphi: M_g \to N$ a harmonic map into a Kähler manifold. If $r_\varphi > 2g-2$ then ψ' and ψ'' are J_2-holomorphic. In particular, if $M = S^2$ then φ is harmonic if and only if ψ' and ψ'' are J_2-holomorphic.*

Proof. There is only anything to prove when φ is non-constant and then we have the line bundle L_φ spanned locally by δ. The inequality on the ramification is just the condition that $\deg L_\varphi \geq 1$. Consider the projection of L_φ into T/T_1. Since $\sup T/T_1 < 1$ and $\inf L_\varphi = \deg L_\varphi \geq 1$ the projection must be zero and hence $L_\varphi \subset T_1$. Since $T_1 \subset \underline{\psi} \cap \underline{\psi}'$ the proof is complete. □

In fact this result can be improved as follows: we split the differential of φ into (1,0) and (0,1) parts with respect to N. That is, let $\delta = \alpha + \bar{\beta}$, then φ is holomorphic if and only if $\bar{\beta}$ vanishes identically. Thus if φ is neither holomorphic nor anti-holomorphic then both $dz \otimes \alpha$ and $dz \otimes \bar{\beta}$ are holomorphic and non-zero so have well-defined orders of vanishing $r_\varphi{}'$ and $r_\varphi{}''$ as well as determining subbundles $A, \bar{B}$ of T', T'' respectively. Clearly

$$r_\varphi' = c_1(\kappa_M \otimes A), \quad r_\varphi'' = c_1(\kappa_M \otimes \bar{B})$$

as above so $A \subset T'_k$, $\bar{B} \subset T''_l$ where $k \le r_\varphi' + \chi(M)$ and $l \le r_\varphi'' + \chi(M)$. Thus if $r_\varphi' + r_\varphi'' + 2\chi(M) \ge 1$ there is an integer k with $A + \bar{B} \subset T'_k + T''_{1-k}$ which by lemma (2.9) is sufficient to give us a J_2-holomorphic lift ψ since $L_\varphi \subset A + \bar{B}$. Thus we have shown

Theorem 2.11. *If the holomorphic and antiholomorphic ramifications r_φ', r_φ'' of a non $\pm$holomorphic harmonic map $\varphi: M_g \to N$ of a Riemann surface of genus g into a Kähler manifold are defined as above and satisfy*

$$r_\varphi' + r_\varphi'' \ge 4g - 3$$

then, for all k with $2g-1-r_\varphi'' \le k \le r_\varphi' - 2g - 2$, $\underline{\psi} = T'_k + T''_{1-k}$ defines a J_2-holomorphic lift.

Remarks. (a) This result strengthens theorem (2.10) since $r_\varphi \le \frac{1}{2}(r_\varphi' + r_\varphi'')$.

(b) Eells-Salamon [33] have also provided lifts into $J(N)$ for harmonic maps of a Riemann surface M into Kähler manifolds N using a Gauss map $M \to P(T'N) \hookrightarrow J(N)$. Their lift requires no restriction on r_φ but our lifts are easier to use since $\underline{\psi}'$ and $\underline{\psi}''$ are composed of semistable bundles of positive slope.

Let us now turn to the case of an arbitrary target manifold N. When M has genus > 0 we have no way to extend T_1 to a maximal isotropic subbundle, so our method only produces lifts to bundles of f-structures (c.f. [61]). However when $M = S^2$ we can do better since we have available a considerable refinement of the HN filtration due to Birkhoff-Grothendieck:

Theorem 2.12. [40] *Let $E \to S^2$ be a holomorphic vector bundle. Then there is a decomposition of E as a sum of line bundles*

$$E = L_1 + L_2 + \ldots + L_k$$

with $\deg L_i \ge \deg L_{i-1}$.

From this we see that a vector bundle over S^2 is semi-stable if and only if it is a sum of line bundles all of the same degree. Further in the HN filtration of a general bundle, E_i/E_{i+1} is a direct sum of copies of the line bundle of degree i. In this case we can characterise E_i as the sum of all line subbundles of E of degree $\ge i$ or the span of all meromorphic sections of E whose divisors have degree $\ge i$.

From theorem (2.12), T_0/T_1 is holomorphically trivial so pointwise evaluation of sections identifies each fibre with the space $\mathrm{H}^0(T_0/T_1)$ of holomorphic sections. Since T_1 is the polar of T_0 the quotient T_0/T_1 has an induced nondegenerate bilinear form which gives one also to $\mathrm{H}^0(T_0/T_1)$. Choosing any isotropic subspace of $\mathrm{H}^0(T_0/T_1)$ gives an isotropic subbundle of T_0 containing T_1. If N is even dimensional so is $\mathrm{H}^0(T_0/T_1)$ and so we can always find a maximal isotropic subbundle of T containing T_1. Arguing as in theorem (2.10) we see we have constructed a J_2-holomorphic map of S^2 into $J(N)$ covering φ.

Theorem 2.13. *Any harmonic map $\varphi: S^2 \to N^{2n}$ has a J_2-holomorphic lift $\psi: S^2 \to J(N)$ which may be constructed by choosing a maximal isotropic subspace of $\mathrm{H}^0(T_0/T_1)$ or equivalently by choosing sufficiently many holomorphic sections of $\varphi^{-1}TN^{\mathbf{C}}$.*

Remark. Theorem (2.5) proved the existence of a J_2-holomorphic lift for all Riemann surfaces M without any assumption on the ramification. The advantage of the construction above over theorem (2.5) is that it gives an explicit recipe for constructing ψ.

The almost-complex manifolds $(J(N),J_1)$, $(J(N),J_2)$ are in general very badly behaved: for instance J_2 is never integrable. Concerning J_1 we have:

Proposition 2.14. [55] *Let $j \in J(N)$ with $\sqrt{-1}$-eigenspace $T^+ \subset T_{\pi(j)}N^{\mathbf{C}}$. Let R denote the Riemann curvature tensor of ∇^N. Then the Nijenhuis tensor of J_1 vanishes at j if and only if*

$$R(T^+,T^+)T^+ \subset T^+ .$$

It follows from proposition (2.14) that J_1 is integrable only when N is conformally flat (see, for example, [55, 30]). Further, there only exist Hermitian metrics on $J(N)$ with good compatibility properties with J_1 or J_2 under very stringent curvature conditions, (c.f. Rawnsley [61], Hitchin [43]). The relative pathology of J_1 and J_2 leads us to seek better behaved twistor spaces that reflect the geometry of N. This programme can be successfully carried out when N is a symmetric space as we shall see below.

Chapter 3. Symmetric Spaces

The most important class of homogeneous spaces is that of Riemannian symmetric spaces. The remainder of this monograph is devoted to the study and applications of twistor spaces over Riemannian symmetric spaces.

In this chapter, therefore, we briefly review the basic facts about such spaces and then discuss their topology. Using the classification theory of Murakami [53], we find a Lie-theoretic method of determining the second homotopy group of a compact Riemannian symmetric space which will prove useful in our study of minimal 2-spheres in Chapter 7.

The basic reference for the properties of Riemannian symmetric spaces is the book of Helgason [42] which the Reader should consult for more details and for definitions and proofs not presented below.

A. Riemannian symmetric spaces

Let G be a Lie group. Recall that a homogeneous G-space is symmetric if it is of the form G/K with $(G^\tau)_0 \subset K \subset G^\tau$ for some involution τ of G. Then $\mathbf{k} = \mathbf{g}^\tau$ and setting $\mathbf{p} = \mathbf{g}^{-\tau}$ gives a reductive summand. We have the following relations:

$$[\mathbf{k}, \mathbf{k}] \subset \mathbf{k} , \qquad [\mathbf{k}, \mathbf{p}] \subset \mathbf{p} , \qquad [\mathbf{p}, \mathbf{p}] \subset \mathbf{k} . \tag{1}$$

Conversely, a decomposition $\mathbf{g} = \mathbf{k} \oplus \mathbf{p}$ satisfying (1) is called a *symmetric decomposition* and gives rise to a family of symmetric spaces.

A symmetric space is *Riemannian* if it admits an invariant metric, for which it is necessary and sufficient that $\mathbf{p}$ admit an $\mathrm{Ad}(K)$-invariant inner product or, equivalently, that $\mathrm{Ad}_{\mathbf{p}}(K)$ be compact.

So let N be a Riemannian symmetric space with Maurer-Cartan form β. From corollary (1.4) and formula (1), we see that the canonical connection of N is torsion-free and so coincides with the Levi-Civita connection of any invariant metric. Further, from lemma (1.5) and formula (1), we see that the Levi-Civita connection ∇ is also given by

$$\beta \circ \nabla = P \circ d \circ \beta , \tag{2}$$

where $P: \underline{\mathbf{g}} \to [\mathbf{p}]$ is projection along $[\mathbf{k}]$. Thus, for Riemannian symmetric spaces, the Levi-Civita connection is just flat differentiation in $\underline{\mathbf{g}}$ followed by projection.

Lastly, again from corollary (1.4), the curvature tensor R of N is given by

$$\beta \circ R = -\tfrac{1}{2}\mathrm{ad}[\beta \wedge \beta]\beta . \tag{3}$$

If a Riemannian symmetric space admits an invariant almost Hermitian structure then it is called a *Hermitian symmetric space*. Since it is invariant, such a structure is parallel for the Levi-Civita connection and so is integrable and Kähler.

B. Types of symmetric space

We distinguish several different types of symmetric spaces according to the algebraic nature of $\mathbf{g}$ and τ.

Symmetric spaces of semi-simple type

Let G/K be a Riemannian symmetric space with involution τ. We say that G/K is *of compact type* if G is compact and semi-simple (equivalently, if the Killing form of $\mathbf{g}$ is negative definite).

Now suppose that $\mathbf{g}$ is non-compact and semi-simple with Killing form B. A *Cartan involution* on $\mathbf{g}$ is an involution τ for which the symmetric form B_τ given by

$$B_\tau(X,Y) = -B(X,\tau Y)$$

is positive definite. The corresponding symmetric decomposition is called a *Cartan decomposition*. It is known that any non-compact semi-simple Lie algebra admits a Cartan involution which is unique up to conjugacy.

G/K is said to be *of non-compact type* if G is non-compact and semi-simple while τ is a Cartan involution. Such symmetric spaces are known to be diffeomorphic to Euclidean spaces.

We shall say that a symmetric space is *of semi-simple type* if it is of compact or non-compact type.

There is a well known duality between non-compact semi-simple Lie algebras and compact semi-simple Lie algebras with involutions: if $\mathbf{g} = \mathbf{k} \oplus \mathbf{p}$ is a Cartan decomposition, then $\mathbf{u} = \mathbf{k} \oplus \sqrt{-1}\mathbf{p}$ is a symmetric decomposition of a compact semi-simple Lie algebra $\mathbf{u}$ and conversely. $\mathbf{u}$ is said to be the *compact dual* of $\mathbf{g}$ while $\mathbf{g}$ is the *non-compact dual* of $\mathbf{u}$. Similarly, a symmetric space U/K of compact type is dual to a symmetric space G/K of non-compact type if the corresponding symmetric decompositions are dual.

The study of Riemannian symmetric spaces essentially reduces to the study of symmetric spaces of semi-simple type by virtue of the following theorem (c.f. [42, p. 244]).

Theorem 3.1. *Let G/K be a simply connected Riemannian symmetric space. Then G/K is a Riemannian product*

$$G/K = N_0 \times N_+ \times N_- \, ,$$

where N_0 is a Euclidean space and N_+ (respectively N_-) is a Riemannian symmetric space of compact (respectively non-compact) type.

These three types may be distinguished by their sectional curvatures. Indeed, denoting the Killing form of $\mathbf{g}$ by B, if G/K is of compact type then $-B$ provides an invariant metric and then, from (3), the sectional curvature is given by

$$\mathrm{Riem}(X,Y) = -B([\beta(X), \beta(Y)],[\beta(X), \beta(Y)]) \geq 0$$

since B is negative definite on $\mathbf{g}$.

Again, if G/K is of non-compact type, B provides an invariant metric on $[\mathbf{p}]$ so that the sectional curvature is given by

$$\mathrm{Riem}(X,Y) = B([\beta(X), \beta(Y)],[\beta(X), \beta(Y)]) \leq 0$$

since B is negative definite on $[\mathbf{k}]$, τ being a Cartan involution.

Of course, the curvature of Euclidean space vanishes identically.

Henceforth, we shall always assume that symmetric spaces of semi-simple type are equipped with the invariant metric provided by $\pm B$.

Amongst the symmetric spaces of semi-simple type, we distinguish those with inner involution. We call these *inner* symmetric spaces. As we shall see below, inner symmetric spaces are always even-dimensional and have a particularly satisfactory twistor theory. It is known that a symmetric space of semi-simple type G/K is inner if and only if $\operatorname{rank} G = \operatorname{rank} K$, that is, if K contains a maximal torus of G (see e.g. [42, p. 424]).

The inner symmetric spaces include the even-dimensional spheres and the Hermitian symmetric spaces of semi-simple type.

It will be of interest in the sequel to determine how many points in a symmetric space of semi-simple type have the same stabiliser. To this end, let $N = G/K$ be a symmetric space of semi-simple type and let Σ_N denote the group $K \backslash N_G(K)$ where $N_G(K)$ is the normaliser of K in G. It is easy to check that Σ_N has a free right action on N by isometries given by

$$gK \cdot Kn = gnK$$

and this action commutes with that of G. Further, Σ_N acts simply transitively on those points of N with stabilizer K.

It is easily seen that the Lie algebra of $N_G(K)$ coincides with that of K so that Σ_N is a discrete group. Thus, if N is of compact type, Σ_N is finite. If N is of non-compact type the situation is even simpler since, by [42, p. 252], there is only one subgroup of G with Lie algebra $\mathbf{k}$ whence Σ_N is trivial.

Examples. (a) If $N = S^{2n} = \mathrm{SO}(2n+1)/\mathrm{SO}(2n)$, then $\Sigma_N = \mathbf{Z}_2$ and the non-trivial element acts as the antipodal map.

(b) We list below those classical irreducible inner symmetric spaces of compact type for which Σ_N is non-trivial:

Type	N	Σ_N
A	$\mathrm{SU}(2n)/\mathrm{S}(\mathrm{U}(n)\times\mathrm{U}(n))$	$\mathbf{Z}_2$
BD	$\mathrm{SO}(n)/\mathrm{SO}(2p)\times\mathrm{SO}(n-2p)$	$\mathbf{Z}_2\times\mathbf{Z}_2$ if $n=4p$ $\mathbf{Z}_2$ if $n\neq 4p$
C	$\mathrm{Sp}(n)/\mathrm{U}(n)$ $\mathrm{Sp}(2p)/\mathrm{Sp}(p)\times\mathrm{Sp}(p)$	$\mathbf{Z}_2$ $\mathbf{Z}_2$
D	$\mathrm{SO}(4n)/\mathrm{U}(2n)$	$\mathbf{Z}_2$

(c) If N is inner, Σ_N may also be described as follows: let W_K be the Weyl group of K viewed as a subgroup of W_G, the Weyl group of G. Let $W_G(K)$ denote the subgroup of elements of W_G that preserve the roots of K. Then one can show that

$$\Sigma_N = W_G(K)/W_K .$$

Using this description, we can check that Σ_N is trivial for $N = \mathrm{G}_2/\mathrm{SO}(4)$.

Hermitian symmetric spaces

The invariant complex structure of a Hermitian symmetric space G/K of semi-simple type arises in a rather simple way. Firstly, observe that an invariant almost Hermitian structure always corresponds to an $\mathrm{Ad}(K)$-invariant maximally isotropic subspace $\mathbf{p}^+ \subset \mathbf{p}^{\mathbf{C}}$ and then the bundle of $(1,0)$ vectors is just $[\mathbf{p}^+]$.

Since $\mathbf{p}^+$ is Ad(K)-invariant, we have

$$[\mathbf{k}, \mathbf{p}^+] \subset \mathbf{p}^+$$

so that

$$B([\mathbf{p}^+, \mathbf{p}^+], \mathbf{k}) = B(\mathbf{p}^+, [\mathbf{k}, \mathbf{p}^+]) = B(\mathbf{p}^+, \mathbf{p}^+) = 0$$

by isotropy of $\mathbf{p}^+$. But, from (1), we have

$$[\mathbf{p}^+, \mathbf{p}^+] \subset \mathbf{k}^{\mathbf{C}}$$

whence $\mathbf{p}^+$ (and its complex conjugate) is abelian.

From this, we can easily check that extending the complex structure on $\mathbf{p}$ by zeros on $\mathbf{k}$ gives a derivation of $\mathbf{g}$. Since $\mathbf{g}$ is semi-simple, this derivation is inner i.e. is just the adjoint action of some element of $\mathbf{g}$ which in this case must lie in the centre of $\mathbf{k}$.

Thus we have proved the well-known

Proposition 3.2. *If N is a Hermitian symmetric space of semi-simple type then the complex structure at $x \in N$ is given by the adjoint action of an element of the centre of the stabiliser of x.*

Irreducible symmetric spaces

A symmetric space G/K of semi-simple type is *irreducible* if ad$\mathbf{k}$ acts irreducibly on $\mathbf{p}$.

Remark. For simply connected symmetric spaces, this amounts to being holonomy irreducible since the metric structure is analytic.

In particular, as a special case of De Rham's theorem we have

Theorem 3.3. *A simply connected Riemannian symmetric space of semi-simple type is a direct product of irreducible symmetric spaces.*

The irreducible symmetric spaces are exhausted by the following four classes:

Type I: $\mathbf{g}$ is compact and simple.

Type II: $\mathbf{g} = \mathbf{g}_1 \oplus \mathbf{g}_1$ with $\mathbf{g}_1$ compact and simple and τ given by

$$\tau(X,Y) = (Y,X) .$$

The corresponding symmetric spaces are the group manifolds G_1 with the symmetric space structure discussed in chapter 1.

Type III: $\mathbf{g}$ is non-compact and simple with no compatible complex structure.

Type IV: $\mathbf{g} = \mathbf{g}_1^{\mathbf{C}}$ where $\mathbf{g}_1$ is compact and simple while τ is just complex conjugation with respect to $\mathbf{g}_1$.

We observe that the first two classes are of compact type and that type I is dual to type III, type II to type IV.

We shall say that a symmetric space is *simple* if it is irreducible of type I or III.

Lastly, let us note that if G/K is of type II or IV then rank $G = 2$ rank K so that all irreducible inner symmetric spaces are simple.

C. Structure theory

We now wish to discuss the classification and topology of Riemannian symmetric spaces. For this, as well as for successive chapters, we require some structure theory of Lie algebras which is collected in this section. For definitions and proofs not presented below, the Reader is referred to the books of Humphreys [46] and Helgason [42].

Root systems

Let $\mathbf{g}^{\mathbf{C}}$ be a complex semi-simple Lie algebra. The Killing form of $\mathbf{g}^{\mathbf{C}}$, denoted B, is an invariant non-degenerate bilinear form.

Let $\mathbf{a}$ be a *Cartan subalgebra* (CSA) of $\mathbf{g}^{\mathbf{C}}$ (thus a maximal set of commuting and semi-simple elements). Given α in the dual space $\mathbf{a}^*$, we put

$$\mathbf{g}^{\alpha} = \{X \in \mathbf{g}^{\mathbf{C}} : [H,X] = \alpha(H), \quad \text{for } H \in \mathbf{a}\}$$

Then $\mathbf{g}^0 = \mathbf{a}$ and the non-zero α with $\mathbf{g}^{\alpha} \neq 0$ are called *roots* with *root spaces* $\mathbf{g}^{\alpha}$. We denote the set of roots by $\Delta(\mathbf{g}^{\mathbf{C}},\mathbf{a})$ or just Δ if there is no ambiguity. The main properties of the root spaces are summarised in

Theorem 3.4. *(i) The decomposition*

$$\mathbf{g}^{\mathbf{C}} = \mathbf{a} + \sum_{\alpha \in \Delta} \mathbf{g}^{\alpha}$$

is a direct sum.

(ii) For each $\alpha \in \Delta$, $\mathbf{g}^{\alpha}$ *is 1-dimensional.*

(iii) For α, $\beta \in \Delta$ *with* $\alpha + \beta \neq 0$, $B(\mathbf{g}^{\alpha}, \mathbf{g}^{\beta}) = 0$.

(iv) B *is non-degenerate on* $\mathbf{a}$ *whence, for* $\alpha \in \mathbf{a}^*$, *there is a unique* $H_{\alpha} \in \mathbf{a}$ *such that*

$$\alpha(H) = B(H_{\alpha}, H)$$

for all $H \in \mathbf{a}$.

(v) If $\alpha \in \Delta$ *then* $-\alpha \in \Delta$ *and for* $X \in \mathbf{g}^{\alpha}$, $Y \in \mathbf{g}^{-\alpha}$ *we have*

$$[X,Y] = B(X,Y)\, H_{\alpha}.$$

(vi) For α, $\beta \in \Delta$, *if* $\alpha + \beta \in \Delta$ *then*

$$[\mathbf{g}^{\alpha}, \mathbf{g}^{\beta}] = \mathbf{g}^{\alpha+\beta} .$$

Now B induces a positive definite inner product on the real span of Δ which we denote by $(.,.)$. The main facts about the inner products of roots are given in the following proposition:

Proposition 3.5. *Let* $\alpha, \beta \in \Delta$. *Then*

(i)

$$2\frac{(\alpha,\beta)}{(\beta,\beta)} \in \mathbf{Z} ,$$

(ii) if $(\beta,\beta) \leq (\alpha,\alpha)$, *then*

$$\frac{(\alpha,\alpha)}{(\beta,\beta)} \in \{1,2,3\} ,$$

(iii) if $(\alpha,\beta) > 0$ *then* $\alpha - \beta \in \Delta$ *while if* $(\alpha,\beta) < 0$ *then* $\alpha + \beta \in \Delta$

Definition. A subset S of Δ is said to be *closed* if whenever $\alpha, \beta \in S$ and $\alpha+\beta \in \Delta$ then $\alpha+\beta \in S$.

Definition. A *positive root system* is a subset Δ^+ of Δ such that

(i) $\Delta^+ \cap -\Delta^+ = \varnothing$;
(ii) Δ^+ is closed;
(iii) $\Delta^+ \cup -\Delta^+ = \Delta$.

The elements of Δ^+ are called *positive roots.*

As is well known, a choice of positive root system is equivalent to a choice of Weyl chamber or a lexicographic ordering on the real span of Δ.

Remark. Any subset $\Phi \subset \Delta$ satisfying (i) and (ii) can be extended to a positive root system [8].

Definition. Given a positive root system, a positive root is said to be *simple* if it cannot be written as a sum of two other positive roots.

Proposition 3.6. *Let Δ^+ be a positive root system and $\alpha_1, \ldots, \alpha_l$ be the simple roots. Then*
(i) $\{\alpha_1, \ldots, \alpha_l\}$ are linearly independent;
(ii) if $\alpha \in \Delta^+$ then

$$\alpha = \sum_{i=1}^{l} n_i \alpha_i$$

with each n_i a non-negative integer;
(iii) $l = \dim_{\mathbf{C}} \mathbf{a}$.

The simple roots and their inner products characterise $\mathbf{g}^{\mathbf{C}}$ up to isomorphism.

Finally we note a simple lemma which will be useful below and for which we could not find a proof in the literature.

Lemma 3.7. *Let Δ^+ be a positive root system and $\alpha_1, \ldots, \alpha_l$ be the simple roots. Let $\alpha \in \Delta$ and suppose that*

$$\alpha = \sum_{i=1}^{l} n_i \alpha_i \ .$$

Then $n_i(\alpha_i, \alpha_i)/(\alpha, \alpha) \in \mathbf{Z}$.

Proof. For $\alpha \in \Delta$, set $\hat{\alpha} = 2\alpha/(\alpha,\alpha)$. Then $\{\hat{\alpha} : \alpha \in \Delta\}$ is a root system with simple roots $\{\hat{\alpha}_i\}$ (see [46]). Thus $\hat{\alpha}$ is an integer linear combination of the $\{\hat{\alpha}_i\}$. However, the coefficient of $\hat{\alpha}_i$ in $\hat{\alpha}$ is just $n_i(\alpha_i, \alpha_i)/(\alpha, \alpha)$ and the lemma follows. □

Roots of simple Lie algebras

Now let us fix a CSA $\mathbf{a}$ with roots Δ. Δ is *irreducible* if it cannot be partitioned into two proper non-empty mutually orthogonal subsets.

It is well known that $\mathbf{g}^{\mathbf{C}}$ is simple if and only if the root system Δ is irreducible. As a consequence of this irreducibility we have the following lemmata.

Lemma 3.8. *If $\mathbf{g}^{\mathbf{C}}$ is simple then there are at most two different lengths for $\alpha \in \Delta$. Further, all roots of the same length are conjugate under the Weyl group of $\mathbf{g}^{\mathbf{C}}$.*

Definition. If there are two root lengths in Δ we divide them into *long* and *short* roots. If all roots have the same length we shall say that they are all long.

Now let us choose a positive root system $\Delta^+ \subset \Delta$. We have a partial ordering defined on Δ by setting $\alpha \leq \beta$ if and only if $\beta - \alpha$ is zero or a positive linear combination of simple roots.

Lemma 3.9. *If $\mathbf{g}^{\mathbf{C}}$ is simple and $\Delta^+ \subset \Delta$ is a positive root system then there is a unique element $\theta \in \Delta^+$ which is maximal with respect to the above partial ordering. Furthermore, θ is long.*

We call θ the *highest root* (with respect to Δ^+).

Roots and involutions

We now discuss the interactions between involutions of a compact semi-simple Lie algebra and root systems.

So let $\mathbf{g}$ be such a Lie algebra with involution τ and symmetric decomposition $\mathbf{k} \oplus \mathbf{p}$. Let $(.,.)$ denote the Killing inner product on $\mathbf{g}$.

If $\mathbf{t} \subset \mathbf{g}$ is a maximal torus (i.e a maximal toral subalgebra) then $\mathbf{t}^{\mathbf{C}}$ is a CSA for $\mathbf{g}^{\mathbf{C}}$. Further, for each $X \in \mathbf{g}$, ad$X$ is skew with respect to $(.,.)$ and so has purely imaginary eigenvalues. Thus, for each root $\alpha \in \Delta(\mathbf{g}^{\mathbf{C}}, \mathbf{t}^{\mathbf{C}})$, we have $\alpha \in \sqrt{-1}\mathbf{t}^*$ and

$$\overline{\mathbf{g}^{\alpha}} = \mathbf{g}^{-\alpha} .$$

The following lemma is well known (see e.g. [76]).

Lemma 3.10 *Let $\mathbf{t}_{\mathbf{k}} \subset \mathbf{k}$ be a maximal torus of $\mathbf{k}$. Then $\mathbf{t}_{\mathbf{k}} \oplus \mathbf{z}_{\mathbf{p}}(\mathbf{t}_{\mathbf{k}})$ is a τ-stable maximal torus of $\mathbf{g}^{\mathbf{C}}$.*

Definition. A τ-stable maximal torus of $\mathbf{g}$ whose intersection with $\mathbf{k}$ is maximal abelian in $\mathbf{k}$ is called a *fundamental torus* (with respect to τ).

We fix a fundamental torus $\mathbf{t} = \mathbf{t}_{\mathbf{k}} \oplus \mathbf{t}_{\mathbf{p}}$ and consider the roots $\Delta = \Delta(\mathbf{g}^{\mathbf{C}}, \mathbf{t}^{\mathbf{C}})$ and $\Delta_{\mathbf{k}} = \Delta(\mathbf{k}^{\mathbf{C}}, \mathbf{t}_{\mathbf{k}}^{\mathbf{C}})$. For $\alpha \in \Delta$, we set $\alpha' = \alpha|\mathbf{t}_{\mathbf{k}}$, $\tilde{\alpha} = \alpha \circ \tau$. Since $\mathbf{z}(\mathbf{t}_{\mathbf{k}}) = \mathbf{t}$, we see that α' is non-zero for all roots α. We partition Δ by setting

$$I = \{\alpha \in \Delta: \ \alpha|\mathbf{t}_{\mathbf{p}} = 0\} \qquad II = \{\alpha \in \Delta: \ \alpha|\mathbf{t}_{\mathbf{p}} \neq 0\}$$

so that $\Delta = I \cup II$. Further, $\alpha \in I$ if and only if $\alpha = \tilde{\alpha}$. This happens precisely when $\mathbf{g}^{\alpha} \subset \mathbf{k}^{\mathbf{C}}$ or $\mathbf{g}^{\alpha} \subset \mathbf{p}^{\mathbf{C}}$. In the first case we say $\alpha \in I_{\mathbf{k}}$ and in the second $\alpha \in I_{\mathbf{p}}$. Thus $\Delta = I_{\mathbf{k}} \cup I_{\mathbf{p}} \cup II$.

Remark. If τ is inner then $\mathbf{t}_{\mathbf{k}}$ is maximal abelian in $\mathbf{g}$ and so $\mathbf{t}_{\mathbf{p}}$ is zero and $\Delta = I_{\mathbf{k}} \cup I_{\mathbf{p}}$.

Lemma 3.11. $\Delta_{\mathbf{k}} = \{\alpha': \ \alpha \in I_{\mathbf{k}} \cup II\}$.

Proof. Clearly $\Delta_{\mathbf{k}} \subset \{\alpha': \ \alpha \in I_{\mathbf{k}} \cup II\}$ and if $\alpha \in I_{\mathbf{k}}$, then $\alpha = \alpha' \in \Delta_{\mathbf{k}}$. It remains to show that $\beta' \in \Delta_{\mathbf{k}}$ for $\beta \in II$. For this, let $x_\beta + y_\beta \in \mathbf{g}^\beta$ with $x_\beta \in \mathbf{k}^{\mathbf{C}}$ and $y_\beta \in \mathbf{p}^{\mathbf{C}}$. Then, for $\xi \in \mathbf{t}_{\mathbf{k}}$,

$$[\xi, x_\beta + y_\beta] = \beta(\xi) x_\beta + \beta(\xi) y_\beta$$

and since adξ preserves the symmetric decomposition we find that

$$[\xi, x_\beta] = \beta(\xi)x_\beta$$

so that $\beta' \in \Delta_{\mathbf{k}}$ and $x_\beta \in \mathbf{k}^{\beta'}$. □

We may construct a τ-stable positive root system by choosing $\xi \in \sqrt{-1}\mathbf{t}_{\mathbf{k}}$ on which no $\alpha \in \Delta$ vanishes (such ξ lie in the complement of finitely many hyperplanes) and setting $\Delta^+ = \{\alpha \in \Delta:\ \alpha(\xi) > 0\}$. Clearly this defines a τ-stable positive root system and it is readily shown that all such arise in this way. It is now a simple matter to verify

Lemma 3.12. *If Δ^+ is τ-stable then*

$$\Delta_{\mathbf{k}}^+ = \{\alpha' :\ \alpha \in \Delta^+ \cap (I_{\mathbf{k}} \cup II)\}$$

is a positive root system for $\mathbf{k}$. *Further, if* $\mathbf{g}$ *is simple then the highest root* $\theta \in I$.

Finally, if Φ is the set of simple roots for a τ-stable Δ^+, we set $\Phi_{\mathbf{k}} = \Phi \cap I_{\mathbf{k}}$, $\Phi_{\mathbf{p}} = \Phi \cap I_{\mathbf{p}}$, $\Phi_{II} = \Phi \cap II$. These sets are clearly also τ-stable.

D. Topology of symmetric spaces

We now investigate the topology of the symmetric spaces of semi-simple type. Since the symmetric spaces of non-compact type are known to be diffeomorphic to Euclidean spaces, we only consider symmetric spaces of compact type.

In particular, we shall determine the second homotopy groups of the simply connected irreducible symmetric spaces of compact type and study the homotopy classes of certain homogeneous 2-spheres contained therein.

Homotopy and unit lattices

Our approach is based on the following well known proposition (c.f. [69]).

Proposition 3.13 *Let G, G_1 be simply connected Lie groups and $K \subset G$, $K_1 \subset G_1$ closed subgroups. Then $\pi_2(G/K) \cong \pi_1(K)$ and the isomorphism is functorial in that if $\rho: G \to G_1$ is a homomorphism and $\varphi: G/K \to G_1/K_1$ a ρ-equivariant base-point preserving map then the following diagram commutes:*

$$\begin{array}{ccc} \pi_2(G/K) & \cong & \pi_1(K) \\ \downarrow \varphi_* & & \downarrow \rho_* \\ \pi_2(G_1/K_1) & \cong & \pi_1(K_1) \end{array}$$

Proof. Since G is a Lie group, $\pi_2(G)$ vanishes. Thus the long exact homotopy sequence for the fibration $K \to G \to G/K$ gives

$$0 \to \pi_2(G/K) \to \pi_1(K) \to 0\ .$$

The last assertion now follows from the functoriality of the long exact sequence. □

We now wish to determine $\pi_1(K)$. For this let $T \subset K$ be a maximal torus with Lie algebra $\mathbf{t}$ and K_{ss} the semi-simple (hence closed) subgroup of K with Lie algebra $[\mathbf{k}, \mathbf{k}]$. Lastly let $\tilde{K}_{ss}$ be the simply connected covering group of K_{ss}.

We introduce lattices Γ, Λ as follows:

$$\Gamma = \{\xi \in \mathbf{t}:\ \exp_K 2\pi\xi = e\}\ ,$$

$$\Lambda = \{\xi \in \mathbf{t} \cap [\mathbf{k},\mathbf{k}]:\ \exp_{\tilde{K}_{ss}} 2\pi\xi = e\}\ .$$

For $\xi \in \Gamma$, define a loop $\gamma(\xi): S^1 \to K$ by $\gamma(\xi)(e^{\sqrt{-1}t}) = \exp_K 2\pi t\xi$ and a map $\Gamma \to \pi_1(K)$ by $\xi \mapsto [\gamma(\xi)]$. Clearly, if $\xi \in \Lambda$ then $\gamma(\xi)$ is null-homotopic in K_{ss} and hence in K. Thus we have a well-defined homomorphism $\eta: \Gamma/\Lambda \to \pi_1(K)$.

Proposition 3.14. *For any compact connected Lie group K and maximal torus $T \subset K$, the homomorphism $\eta: \Gamma/\Lambda \to \pi_1(K)$ is an isomorphism.*

Proof. If K is semi-simple, the result is known (see e.g. [75]).

For the general case, we first observe that since any homotopy class can be represented by a closed geodesic, the map $\Gamma \to \pi_1(T)$ given by $\xi \mapsto [\gamma(\xi)]$ is surjective. Further, it is known that $\pi_1(K/T)$ is trivial ([42, p. 307]) so that the long exact sequence for $T \to K \to K/T$ gives the surjectivity of the natural map $\pi_1(T) \to \pi_1(K)$ and hence of η.

To show that η is injective requires some machinery. Let $\mathbf{z} \subset \mathbf{k}$ be the centre of $\mathbf{k}$ so that

$$\mathbf{k} = \mathbf{z} \oplus [\mathbf{k},\mathbf{k}]$$

and let $z: \mathbf{k} \to \mathbf{z}$ be the corresponding projection. Let Θ be the Maurer-Cartan form of K. Following [44], we introduce a homomorphism $\rho: \pi_1(K) \to \mathbf{z}$ by

$$\rho([\gamma]) = \frac{1}{2\pi}\int_\gamma z \circ \Theta\ .$$

From the structure equations we see that $z \circ \Theta$ is closed so that ρ is well-defined and an easy calculation gives

$$\rho([\gamma(\xi)]) = z(\xi)\ .$$

Thus, if $\xi + \Lambda \in \ker\eta$ we conclude that $\xi \in \mathbf{t} \cap [\mathbf{k},\mathbf{k}]$ whence $\gamma(\xi)$ is a loop in K_{ss}. Consider now the long exact sequence for $K_{ss} \to K \to K/K_{ss}$. Since K/K_{ss} is a torus and hence has vanishing π_2 we conclude that $\pi_1(K_{ss})$ injects into $\pi_1(K)$. Thus our result follows from that for K_{ss}. □

To identify the lattices Γ and Λ, we require some notation. Let $\mathbf{g}$ be a compact semi-simple Lie algebra with maximal torus $\mathbf{t}$ and Killing inner product $(.,.)$. For $\lambda \in (\mathbf{t}^{\mathbf{C}})^*$, define $H_\lambda \in \mathbf{t}^{\mathbf{C}}$ by

$$(H_\lambda, \xi) = \lambda(\xi)$$

for $\xi \in \mathbf{t}^{\mathbf{C}}$. Now, if $\alpha \in \Delta = \Delta(\mathbf{g}^{\mathbf{C}}, \mathbf{t}^{\mathbf{C}})$, define $\eta_\alpha \in \mathbf{t}$ by

$$\eta_\alpha = 2\sqrt{-1}H_\alpha/(\alpha,\alpha)$$

and put $\Lambda_{\mathbf{g}}(\mathbf{t}) = \mathrm{span}_{\mathbf{Z}}\{\eta_\alpha : \alpha \in \Delta\}$. It can be shown that if $\Phi \subset \Delta$ is the set of simple roots for some Weyl chamber then $\{\eta_\alpha : \alpha \in \Phi\}$ is a basis for $\Lambda_{\mathbf{g}}(\mathbf{t})$. From [42, p. 317], we have

Proposition 3.15. *Let G be a compact simply connected semi-simple Lie group and $T \subset G$ a maximal torus. Then*

$$\{\xi \in \mathbf{t}:\ \exp_G 2\pi\xi = e\} = \Lambda_{\mathbf{g}}(\mathbf{t})\ .$$

Putting all this together, we finally conclude

Theorem 3.16. *Let G/K be a simply connected symmetric space of compact type and $\mathbf{t}=\mathbf{t_k}\oplus\mathbf{t_p}$ a fundamental torus. Then*

$$\pi_2(G/K)\cong\Lambda_{\mathbf{g}}(\mathbf{t})\cap\mathbf{k}/\Lambda_{[\mathbf{k},\mathbf{k}]}(\mathbf{t_k}\cap[\mathbf{k},\mathbf{k}]) .$$

Proof. Without loss of generality, we may suppose that G is simply connected and then from (3.15)

$$\Gamma = \{\xi\in\mathbf{t_k}:\ \exp_K 2\pi\xi = e\} = \{\xi\in\mathbf{t_k}:\ \exp_G 2\pi\xi = e\} = \Lambda_{\mathbf{g}}(\mathbf{t})\cap\mathbf{k}$$

while

$$\Lambda = \{\xi\in\mathbf{t_k}\cap[\mathbf{k},\mathbf{k}]:\ \exp_{K_{ss}} 2\pi\xi = e\} = \Lambda_{[\mathbf{k},\mathbf{k}]}(\mathbf{t_k}\cap[\mathbf{k},\mathbf{k}]) .$$

The result now follows from (3.13) and (3.14). □

Remark. In the case where K is semi-simple this theorem was proved by Takeuchi [69] who used it with a case by case analysis to calculate π_2 of the compact irreducible symmetric spaces. We shall use it below to provide a simple root-theoretic method for determining π_2 of such a space.

π_2 of compact irreducible symmetric spaces

In the last section we saw that finding the second homotopy group of a compact symmetric space of semi-simple type G/K reduces to the algebraic problem of determining the quotient of the lattices $\Lambda_{\mathbf{g}}\cap\mathbf{k}$ and $\Lambda_{[\mathbf{k},\mathbf{k}]}$. Further, by theorem (3.3) we may assume that G/K is irreducible and, indeed, is of type I since the type II symmetric spaces are just the group manifolds of the compact simple Lie groups which all have trivial π_2. Thus we shall assume that G is compact and simple.

To proceed we need to know the involutions of a compact simple Lie algebra up to conjugacy. There are several approaches to this problem (see, for example, [25, 42, 53]) but we shall follow Murakami [53].

So let $\mathbf{g}$ be a compact simple Lie algebra. We fix, once and for all, a maximal torus $\mathbf{t}\subset\mathbf{g}$ and a positive root system $\Delta^+\subset\Delta(\mathbf{g}^{\mathbf{C}},\mathbf{t}^{\mathbf{C}})$. We denote the set of simple roots of Δ^+ by Φ and the highest root by θ.

We now distinguish two cases:

(i) inner involutions

Let $\Phi=\{\alpha_1,\ldots,\alpha_l\}$ and define $\xi_1,\ldots,\xi_l\in\mathbf{t}$ by

$$\alpha_i(\xi_j) = \sqrt{-1}\delta_{ij} .$$

Further, define $m_1,\ldots,m_l\in\mathbf{Z}^+$ by

$$\theta = \sum_i m_i\alpha_i .$$

The conjugacy problem for inner involutions is now settled by the following theorem of Borel-de Siebenthal [9], (c.f. [53]).

Theorem 3.17. *If $m_i=1$ or 2, define an involution τ_i by*

$$\tau_i = \mathrm{Ad}\exp\pi\xi_i .$$

Then any inner involution is conjugate to some τ_i. Further, if $\mathbf{k}_i$ is the fixed set of τ_i then $\mathbf{t}$ is a *maximal torus of* $\mathbf{k}_i$ *and*

i) if $m_i = 1$, $\mathbf{k}_i$ *has a one dimensional centre spanned by* ξ_i *while* $\{\alpha_j\}_{j \neq i}$ *is a set of simple roots for the semi-simple part of* $\mathbf{k}_i$,

ii) if $m_i = 2$, $\mathbf{k}_i$ *is semi-simple with a set of simple roots given by* $\{\alpha_j\}_{j \neq i} \cup \{-\theta\}$.

Turning now to the lattices, we see that $\Lambda_{\mathbf{g}}(\mathbf{t}) \cap \mathbf{k}_i = \Lambda_{\mathbf{g}}(\mathbf{t})$. We can now prove

Proposition 3.18. *Let* τ_i *be defined as in (3.17) and set* $\mathbf{k} = \mathbf{k}_i$. *Then*

i) if $m_i = 1$, α_i *is long and* $\Lambda_{\mathbf{g}} / \Lambda_{[\mathbf{k},\mathbf{k}]} \cong \mathbf{Z}$ *with generator* $[\eta_{\alpha_i}]$. *Under this isomorphism, for* $\alpha \in I_{\mathbf{p}} \cap \Delta^+$,

$$[\eta_\alpha] = \frac{(\alpha_i, \alpha_i)}{(\alpha, \alpha)} \in \mathbf{Z} ,$$

ii) if $m_i = 2$ *and* α_i *is short then* $\Lambda_{\mathbf{g}} = \Lambda_{\mathbf{k}}$ *and all roots in* $I_{\mathbf{p}}$ *are short,*

iii) if $m_i = 2$ *and* α_i *is long then* $\Lambda_{\mathbf{g}} / \Lambda_{\mathbf{k}} \cong \mathbf{Z}_2$ *with generator* $[\eta_{\alpha_i}]$. *Under this isomorphism, for* $\alpha \in I_{\mathbf{p}} \cap \Delta^+$,

$$[\eta_\alpha] = \frac{(\alpha_i, \alpha_i)}{(\alpha, \alpha)} \mod 2 .$$

Proof. $\Lambda_{\mathbf{g}}(\mathbf{t})$ has $\mathbf{Z}$-basis $\{\eta_{\alpha_j}\}$ and we check that, for $\alpha \in \Delta$, if $\alpha = \sum n_j \alpha_j$ then

$$\eta_\alpha = \sum n_j \frac{(\alpha_j, \alpha_j)}{(\alpha, \alpha)} \eta_{\alpha_j} . \qquad (4)$$

We also see from the definition of τ_i that $\alpha \in I_{\mathbf{p}} \cap \Delta^+$ precisely when $n_i = 1$.

If $m_i = 1$, we conclude from lemma (3.7) that α_i has the same length as θ and so is long. Further, by theorem (3.17), a $\mathbf{Z}$-basis for $\Lambda_{[\mathbf{k},\mathbf{k}]}$ is given by $\{\eta_{\alpha_j}\}_{i \neq j}$. Thus we see that $\Lambda_{\mathbf{g}} / \Lambda_{[\mathbf{k},\mathbf{k}]} \cong \mathbf{Z}$ with generator $[\eta_{\alpha_i}]$ and that

$$[\eta_\alpha] = \frac{(\alpha_i, \alpha_i)}{(\alpha, \alpha)} [\eta_{\alpha_i}]$$

for all $\alpha \in I_{\mathbf{p}} \cap \Delta^+$.

Now suppose that $m_i = 2$. Again we conclude from lemma (3.7) that $(\alpha_i, \alpha_i)/(\theta, \theta)$ is unity or $\frac{1}{2}$ according as α_i is long or short. Further, by theorem (3.17), a $\mathbf{Z}$-basis for $\Lambda_{\mathbf{k}}$ is given by $\{\eta_{\alpha_j}\}_{i \neq j} \cup \{-\eta_\theta\}$. We have from (4)

$$\eta_\theta = 2 \frac{(\alpha_i, \alpha_i)}{(\theta, \theta)} \eta_{\alpha_i} + \sum_{i \neq j} m_j \frac{(\alpha_j, \alpha_j)}{(\theta, \theta)} \eta_{\alpha_j} .$$

Thus when α_i is short, $\Lambda_{\mathbf{g}} = \Lambda_{[\mathbf{k},\mathbf{k}]}$ and, for $\alpha \in I_{\mathbf{p}} \cap \Delta^+$, the coefficient of η_{α_i} in η_α is $(\alpha_i, \alpha_i)/(\alpha, \alpha)$ which is an integer so that α is also short.

Lastly, if α_i is long, $\eta_{\alpha_i} \notin \Lambda_{[\mathbf{k},\mathbf{k}]}$ but $2\eta_{\alpha_i} \in \Lambda_{[\mathbf{k},\mathbf{k}]}$ so that $\Lambda_{\mathbf{g}} / \Lambda_{[\mathbf{k},\mathbf{k}]} \cong \mathbf{Z}_2$ with generator $[\eta_{\alpha_i}]$ and

$$[\eta_\alpha] = [\frac{(\alpha_i, \alpha_i)}{(\alpha, \alpha)} \eta_{\alpha_i}] = \frac{(\alpha_i, \alpha_i)}{(\alpha, \alpha)} \mod 2$$

for all $\alpha \in I_{\mathbf{p}}$. □

Remarks. *(i)* If $m_i = 1$, $\mathrm{ad}\xi_i$ has eigenvalues $\pm\sqrt{-1}$ on $\mathbf{p}$ and so the corresponding symmetric space is Hermitian. Indeed, all compact simple Hermitian symmetric spaces arise in this way since if $m_i = 2$, $\mathbf{k}_i$ is semi-simple and hence has no centre. Thus we recover the well-known result that $\pi_2 = \mathbf{Z}$ for a compact simple Hermitian symmetric space.

(ii) It is clear from the above development that the various τ_i and $\mathbf{k}_i$ can be read off the extended Dynkin diagram of $\mathbf{g}^{\mathbf{C}}$ as can π_2 of the corresponding symmetric space since that depends solely on the length of α_i and the integer m_i.

Finally, from (3.16) and (3.18) we draw the following

Corollary 3.19. *Let G/K be a compact simply connected inner symmetric space. Then $\pi_2(G/K) = 0$ if and only if, for any maximal torus of K, all $I_{\mathbf{p}}$ roots are short.*

In particular, for $G = \mathbf{SU}(n)$, $\mathbf{SO}(2n)$, $\mathbf{E}_6$, $\mathbf{E}_7$ or $\mathbf{E}_8$, any inner symmetric G-space has non-trivial π_2 since for these groups all roots are long.

(ii) non-inner involutions

The situation for non-inner involutions is rather more complicated. First, we note that the existence of non-inner involutions of $\mathbf{g}$ corresponds with the existence of non-trivial involutions of the Dynkin diagram of $\mathbf{g}^{\mathbf{C}}$. More precisely, let Aut(g) denote the group of automorphisms of $\mathbf{g}$ and Inn(g) the normal subgroup of inner automorphisms. Further, let $\mathcal{S}$ denote the group of orthogonal permutations of the simple roots Φ, i.e the automorphisms of the Dynkin diagram. It is known that $\mathcal{S}$ is isomorphic to Aut(g)/Inn(g). Since the Dynkin diagram of a simple Lie algebra has at most one multi-link, we deduce that if $\mathbf{g}$ admits a non-inner involution then all roots of $\mathbf{g}$ are long. Indeed, we see by inspection of Dynkin diagrams that $\mathbf{a}_n$, $\mathbf{d}_n$ and $\mathbf{e}_6$ are the only simple Lie algebras admitting non-inner involutions.

Now let $\sigma\in\mathcal{S}$ be an involution. We define a representative of the corresponding coset in Aut(g)/Inn(g) as follows. Extend σ by linearity to give an involution of $\sqrt{-1}\mathbf{t}^*$ and hence, by duality, of $\mathbf{t}$. Call this involution τ_σ. Now fix root vectors $e_\alpha\in\mathbf{g}^\alpha$ for $\alpha\in\Phi$ and extend τ_σ to the span of these vectors by

$$\tau_\sigma(e_\alpha) = e_{\sigma(\alpha)} \quad .$$

It is known that τ_σ has a unique extension to a non-inner involution of $\mathbf{g}$ (see [46, p. 75]).

Proposition 3.20. *Let $\sigma\in\mathcal{S}$ and τ_σ be defined as above. Let $\mathbf{k}$ be the fixed set of τ_σ. Then*

i) $\mathbf{t}$ is a fundamental torus for τ_σ and Δ^+ is τ_σ-stable.

ii) $\Phi\subset I_{\mathbf{k}}\cup II$.

iii) $\mathbf{k}$ is simple and the set of simple roots for $\Delta^+_{\mathbf{k}}$ is given by $\{\alpha' : \alpha\in\Phi\}$.

Proof. Parts *i*) and *ii*) are due to Murakami [53] as is the fact that $\mathbf{k}$ is semi-simple and has the asserted simple roots. It remains to check that $\mathbf{k}$ is simple. If $\mathbf{k}$ were not simple then there is a non-trivial partition of the simple roots of $\mathbf{k}$ into mutually orthogonal disjoint subsets $\Phi_1{}'$ and $\Phi_2{}'$. Define Φ_i by

$$\Phi_i = \{\alpha\in\Phi : \alpha'\in\Phi_i{}'\} \ .$$

If $\alpha\in\Phi_1$, $\beta\in\Phi_2$, then

$$0 = (\alpha',\beta') = \tfrac{1}{2}(\alpha,\beta)+\tfrac{1}{2}(\alpha,\tilde{\beta})$$

and since α, β and $\tilde{\beta}$ are all simple roots, both these last summands are non-positive by (3.5iii) and hence vanish. Thus the Φ_i are orthogonal contradicting the irreducibility of Δ. □

Lemma 3.21. *Let τ_σ, $\mathbf{k}$ be as above and $\alpha \in I$. Then α is conjugate to the highest root θ under the Weyl group of* $\mathbf{k}$.

Proof. The adjoint action of $\mathbf{g}$ restricts to give representations of $\mathbf{k}$ on both $\mathbf{k}$ and $\mathbf{p}$. The weights of these representations are just the restrictions of the roots Δ to $\mathbf{t}_{\mathbf{k}}^{\mathbf{C}}$ and since, for $\alpha \in I$, $\beta \in II$, $(\alpha',\alpha') > (\beta',\beta')$, we see that the Weyl group of $\mathbf{k}$ preserves I. Now any weight is Weyl($\mathbf{k}$) conjugate to a dominant weight and further, if $\alpha \in I$ is dominant in $\mathbf{k}$ then by (3.20iii) we have, for all $\beta \in \Phi$,

$$(\alpha,\beta) = (\alpha,\beta') \geq 0 \ ,$$

whence α is dominant in $\mathbf{g}$ and so coincides with θ which is the unique long dominant root. □

Since Weyl($\mathbf{k}$) also preserves the decomposition $I = I_{\mathbf{k}} \cup I_{\mathbf{p}}$, we deduce from (3.21):

Corollary 3.22. *With respect to τ_σ, precisely one of $I_{\mathbf{k}}$, $I_{\mathbf{p}}$ is non-empty.*

Let us now turn to the lattices. For any non-inner involution, if $\Phi_I = \{\alpha_1,\dots,\alpha_r\}$, $\Phi_{II} = \{\beta_1,\dots,\beta_s,\tilde{\beta}_1,\dots,\tilde{\beta}_s\}$ then a $\mathbf{Z}$-basis for $\Lambda_{\mathbf{g}} \cap \mathbf{k}$ is given by $\{\eta_{\alpha_i}\} \cup \{\eta_{\beta_j} + \eta_{\tilde{\beta}_j}\}$. Further, we see that for $\beta \in II$,

$$\eta_{\beta'} = \tfrac{1}{2}\frac{(\beta,\beta)}{(\beta',\beta')}(\eta_\beta + \eta_{\tilde{\beta}}) \ . \tag{5}$$

Concerning the ratio in (5) we have

Lemma 3.23. *If $\beta \in \Phi_{II}$ then*

$$\frac{(\beta',\beta')}{(\beta,\beta)} = \begin{cases} \tfrac{1}{2} & \text{if } (\beta,\tilde{\beta}) = 0, \\ \tfrac{1}{4} & \text{otherwise.} \end{cases}$$

Proof. Since all roots of $\mathbf{g}^{\mathbf{C}}$ are long and β is simple we conclude that $2(\beta,\tilde{\beta})/(\beta,\beta)$ is zero or -1. The result follows now from a trivial calculation using $(\beta',\beta') = \tfrac{1}{2}(\beta,\beta) + \tfrac{1}{2}(\beta,\tilde{\beta})$. □

Proposition 3.24 *Let τ_σ be as above and $\Phi = \Phi_{\mathbf{k}} \cup \Phi_{II}$ the decomposition of the simple roots.*

(i) If $\Phi_{\mathbf{k}}$ is empty then $I_{\mathbf{k}}$ is empty, $\Lambda_{\mathbf{g}} \cap \mathbf{k}/\Lambda_{\mathbf{k}} \cong \mathbf{Z}_2$ and, under this isomorphism, for $\alpha \in I_{\mathbf{p}}$,

$$[\eta_\alpha] = 1 \ .$$

(ii) If $\Phi_{\mathbf{k}}$ is non-empty then $I_{\mathbf{p}}$ is empty and $\Lambda_{\mathbf{g}} \cap \mathbf{k} = \Lambda_{\mathbf{k}}$.

Proof. Firstly, from proposition (3.20), we see that a $\mathbf{Z}$-basis for $\Lambda_{\mathbf{k}}$ is given by $\{\eta_{\alpha_i}\} \cup \{\eta_{\beta_j'}\}$. Now if $\Phi_{\mathbf{k}}$ is non-empty, we see from (3.22) that $I_{\mathbf{p}}$ is empty while, by (3.23) and (3.5ii), if $\beta_j \in \Phi_{II}$ and $\alpha \in I$, then $(\alpha,\alpha) = 2(\beta_j',\beta_j')$. Thus $(\beta_j,\beta_j) = 2(\beta_j',\beta_j')$ for all j so that from (5) we have

$$\eta_{\beta_j'} = \eta_{\beta_j} + \eta_{\tilde{\beta}_j}$$

whence $\Lambda_{\mathbf{g}}\cap\mathbf{k}=\Lambda_{\mathbf{k}}$. This settles (ii).

If, on the other hand, $\Phi_{\mathbf{k}}$ is empty, then σ is an involution of the Dynkin diagram without fixed points and by inspection of the diagrams we conclude that $\mathbf{g}=\mathbf{su}(2n+1)$ and that there is a unique $\beta_i\in\Phi_{II}$ for which $(\beta_i,\tilde{\beta}_i)$ is non-zero. Thus, by (3.23), we have

$$(\beta_i,\beta_i)/(\beta_i',\beta_i') = 4, \qquad (\beta_j,\beta_j)/(\beta_j',\beta_j') = 2 \ \text{ for } i\neq j\ .$$

In particular, since any $I_{\mathbf{k}}$ root would restrict to a root of $\mathbf{k}$ with squared length four times that of β_i' contradicting (3.5ii), we conclude that $I_{\mathbf{k}}$ is empty. We now see from (5) that

$$\eta_{\beta_i}+\eta_{\tilde{\beta}_i} = \tfrac{1}{2}\eta_{\beta_i'}, \qquad \eta_{\beta_j}+\eta_{\tilde{\beta}_j} = \eta_{\beta_j'} \ \text{ for } i\neq j\ ,$$

so that $\Lambda_{\mathbf{g}}\cap\mathbf{k}/\Lambda_{\mathbf{k}}\cong\mathbf{Z}_2$ with generator $[\eta_{\beta_i}+\eta_{\tilde{\beta}_i}]$. It is now straightforward to check that in $\mathbf{su}(2n+1)$, if $\alpha\in I$, $[\eta_\alpha]=[\eta_{\beta_i}+\eta_{\tilde{\beta}_i}]$. □

We can now tackle the conjugacy problem for non-inner involutions which essentially reduces to that for inner involutions on the fixed sets of the τ_σ.

Indeed, for $\sigma\in\mathcal{S}$, we let $\Phi_{\mathbf{k}}=\{\alpha_1,\ldots,\alpha_r\}$, $\Phi_{II}=\{\beta_1,\ldots,\beta_s,\tilde{\beta}_1,\ldots,\tilde{\beta}_s\}$ be the decomposition of Φ with respect to τ_σ and define dual vectors $\xi_1,\ldots,\xi_r\in\mathbf{t}_{\mathbf{k}}$ by

$$\alpha_i(\xi_j) = \delta_{ij}, \qquad \beta_i(\xi_j) = 0\ .$$

Then, if ν is the highest root of $\mathbf{k}$ with respect to $\Delta^+_{\mathbf{k}}$, we write

$$\nu = \sum_i m_i\alpha_i + \sum_j n_j\beta_j'$$

and we have:

Theorem 3.25. [53] *Let $\sigma\in\mathcal{S}$ and $m_i=1$ or 2. Define an involution $\tau_{\sigma,i}$ by*

$$\tau_{\sigma,i} = \tau_\sigma\circ\mathrm{Ad}\exp\pi\xi_i$$

and let $\mathbf{k}_i$ denote its fixed set. Then any non-inner involution of $\mathbf{g}$ is conjugate in Aut(g) *to some τ_σ or $\tau_{\sigma,i}$. Further,*

i) $\mathbf{t}$ is a fundamental torus for $\tau_{\sigma,i}$ and Δ^+ is $\tau_{\sigma,i}$-stable;

ii) the type decomposition of Φ with respect to $\tau_{\sigma,i}$ is given by

$$\Phi_{\mathbf{k}} = \{\alpha_j\}_{j\neq i}, \qquad \Phi_{\mathbf{p}} = \{\alpha_i\}, \qquad \Phi_{II} = \{\beta_1,\ldots,\beta_s,\tilde{\beta}_1,\ldots,\tilde{\beta}_s\}\ ;$$

iii) $\mathbf{k}_i$ is semi-simple and the set of simple roots for $\Delta^+_{\mathbf{k}_i}$ is

$$\{\gamma_i'\}\cup\{\alpha_j'\}_{i\neq j}\cup\{\beta_k'\}\ ,$$

where $\gamma_i\in\Delta^+$ is a sum of successive roots $\alpha_i+\alpha_{i_1}+\ldots+\alpha_{i_k}+\beta_j$ in the Dynkin diagram of $\mathbf{g}^{\mathbf{C}}$ starting at α_i and ending when a type II root is reached.

Remarks. (i) If $\Phi_{\mathbf{k}}$ is empty so that $\mathbf{g}=\mathbf{su}(2n+1)$, we see that no $\tau_{\sigma,i}$ exist and thus τ_σ is the unique non-inner involution up to conjugacy.

(ii) If, on the other hand, $\Phi_{\mathbf{k}}$ is non-empty, it is immediate from corollary (3.22) that $\theta\in I_{\mathbf{k}}$ so that $\nu=\theta$.

Turning to the lattices now we have

Proposition 3.26. *Let $\tau_{\sigma,i}$ be defined as in (3.25) and set $\mathbf{k}=\mathbf{k}_i$. Then $\Lambda_{\mathbf{g}}\cap\mathbf{k}/\Lambda_{\mathbf{k}}\cong\mathbf{Z}_2$ and under this isomorphism, for any $\alpha\in I_{\mathbf{p}}$, $[\eta_\alpha]=1$.*

Proof. From theorem (3.25) we see that a $\mathbf{Z}$-basis for $\Lambda_{\mathbf{k}}$ is given by $\{\eta_{\gamma_i'}\}\cup\{\eta_{\alpha_j}\}_{j\neq i}\cup\{\eta_{\beta_k'}\}$. Further, since $\Phi_{\mathbf{k}}$ is non-empty, we may conclude that

$$\eta_{\beta_k'} = \eta_{\beta_k} + \eta_{\tilde{\beta}_k}$$

by arguing as in (3.24). Now γ_i is of the form $\delta+\beta$ with $\delta\in I,\beta\in\Phi_{II}$ and $\delta-\beta\notin\Delta$ so that by (3.5) we have

$$2\frac{(\delta,\beta)}{(\beta,\beta)} = -1 .$$

Also, we established in the proof of (3.24) that $(\beta,\tilde{\beta})$ vanishes for any $\beta\in\Phi_{II}$. From this we may easily show that $(\gamma_i,\tilde{\gamma}_i)$ vanishes and hence

$$\frac{(\gamma_i',\gamma_i')}{(\gamma_i,\gamma_i)} = \tfrac{1}{2} . \tag{6}$$

Now

$$\eta_{\alpha_i} = \eta_{\gamma_i} - \eta_{\beta_j} + \sum_j \eta_{\alpha_{i_j}}$$

and from (6) we deduce that

$$\eta_{\gamma_i} - \eta_{\beta_j} = \tfrac{1}{2}(\eta_{\gamma_i'} + \eta_{\beta_j'})$$

whence $\eta_{\alpha_i}\notin\Lambda_{\mathbf{k}}$ but $2\eta_{\alpha_i}\in\Lambda_{\mathbf{k}}$. Thus $\Lambda_{\mathbf{g}}\cap\mathbf{k}/\Lambda_{\mathbf{k}}\cong\mathbf{Z}_2$ with generator $[\eta_{\alpha_i}]$. Finally, from the definition of $\tau_{\sigma,i}$ we see that if $\alpha\in I_{\mathbf{p}}$, then α_i has coefficient ±1 in α so that $[\eta_\alpha]=[\eta_{\alpha_i}]$. This concludes the proof. □

We draw an important corollary to the above development. Noting that all roots are of type *II* for a group manifold, we conclude from (3.19), (3.24) and (3.26):

Corollary 3.27. *A simply connected symmetric space of compact type has vanishing π_2 if and only if, with respect to any fundamental torus, all $I_{\mathbf{p}}$ roots are short.*

For G simple, a clear picture has now emerged: a symmetric G-space has $\pi_2=\mathbf{Z}$ precisely when it is Hermitian symmetric and otherwise $\pi_2=\mathbf{Z}_2$ or is trivial depending on whether or not long $I_{\mathbf{p}}$ roots exist.

Example. Consider the exceptional Lie algebra $\mathbf{e}_6$ with Dynkin diagram as shown below:

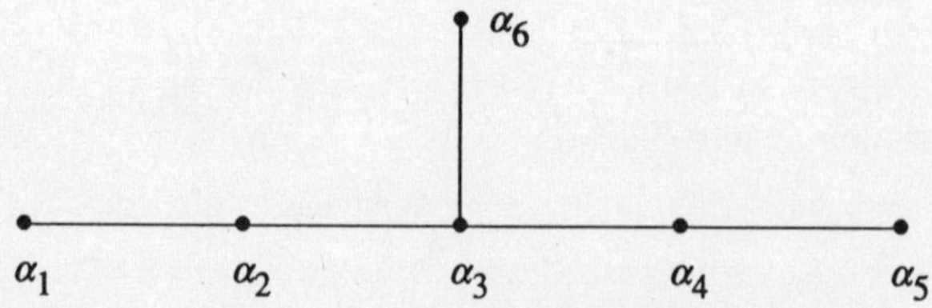

Here $\theta=\alpha_1+2\alpha_2+3\alpha_3+2\alpha_4+\alpha_5+2\alpha_6$. We see that there are four distinct involutions:

corresponding to α_1 is the inner involution with $\mathbf{k}=\mathbf{u}(1)\oplus\mathbf{so}(10)$ and $\pi_2=\mathbf{Z}$; corresponding to α_2 is the inner involution with $\mathbf{k}=\mathbf{su}(2)\oplus\mathbf{su}(6)$ and $\pi_2=\mathbf{Z}_2$. (The inner involutions corresponding to α_5 and those corresponding to α_4 and α_6 are conjugate in $\mathrm{Aut}(\mathbf{e}_6)$ to those for α_1 and α_2 respectively.) The non-inner involution τ_σ has fixed set $\mathbf{f}_4$ and lastly $\tau_{\sigma,6}$ has fixed set $\mathbf{sp}(4)$. Thus the simply connected $\mathbf{E}_6$-symmetric spaces are respectively: $\mathbf{E}_6/\mathbf{U}(1)\times\mathbf{Spin}(10)$, $\mathbf{E}_6/\mathbf{SU}(2)\times_{\mathbf{Z}_2}\mathbf{SU}(6)$, $\mathbf{E}_6/\mathbf{F}_4$ and $\mathbf{E}_6/(\mathbf{Sp}(4)/\mathbf{Z}_2)$.

E. Homogeneous 2-spheres

We now describe some minimal immersions of 2-spheres in symmetric spaces associated with $I_\mathbf{p}$ roots.

Let G/K be a simply connected Riemannian symmetric space of compact type and fix a fundamental torus $\mathbf{t}\subset\mathbf{g}$. If $\alpha\in I_\mathbf{p}$, then $\mathbf{g}^\alpha\oplus\mathbf{g}^{-\alpha}\oplus[\mathbf{g}^\alpha,\mathbf{g}^{-\alpha}]$ is a Lie algebra isomorphic to $\mathbf{sl}(2,\mathbf{C})$ and this is the complexification of a subalgebra $\mathbf{s}_\alpha$ of $\mathbf{g}$ which is isomorphic to $\mathbf{su}(2)$. We let S_α be the corresponding copy of $\mathbf{SU}(2)$ and $U_\alpha\subset S_\alpha$ the connected subgroup with Lie algebra $\mathbf{u}_\alpha=[\mathbf{g}^\alpha,\mathbf{g}^{-\alpha}]\cap\mathbf{g}\subset\mathbf{k}$. Clearly, U_α is a copy of $\mathbf{U}(1)$ so that the homogeneous space S_α/U_α is a 2-sphere.

Now let $i_\alpha:S_\alpha\to G$ be the homomorphism to which the inclusion $\mathbf{s}_\alpha\subset\mathbf{g}$ exponentiates. Then, since $i_\alpha(U_\alpha)\subset K$, we can define an i_α-equivariant immersion $\varphi_\alpha:S^2\to G/K$.

Proposition 3.28. *φ_α is totally geodesic and represents the second homotopy class $[\eta_\alpha]$ of G/K under the isomorphism of theorem (3.16).*

Proof. Put $\mathbf{p}_\alpha=\mathbf{g}\cap(\mathbf{g}^\alpha\oplus\mathbf{g}^{-\alpha})$. Then $\mathbf{s}_\alpha=\mathbf{u}_\alpha\oplus\mathbf{p}_\alpha$ is a symmetric decomposition and $i_\alpha(\mathbf{p}_\alpha)\subset\mathbf{p}$. Thus φ_α is totally geodesic by proposition (1.7).

It is clear from (3.13) that under the isomorphisms of theorem (3.16), the map induced by φ_α on π_2 corresponds with the map $\Lambda_{\mathbf{u}_\alpha}/\Lambda_{[\mathbf{u}_\alpha,\mathbf{u}_\alpha]}\to\Lambda_\mathbf{g}\cap\mathbf{k}/\Lambda_{[\mathbf{k},\mathbf{k}]}$ induced by the inclusion $\mathbf{u}_\alpha\hookrightarrow\mathbf{k}$. Now $\Lambda_{[\mathbf{u}_\alpha,\mathbf{u}_\alpha]}=0$ and $\Lambda_{\mathbf{u}_\alpha}$ is generated by η_α whence φ_α represents $[\eta_\alpha]$. □

Now, from propositions (3.16), (3.24) and (3.26) we find:

Proposition 3.29. *If G/K is a compact simply connected simple symmetric space then if $\pi_2(G/K)$ is non-trivial, it is generated by $[\varphi_\alpha]$ for some $\alpha\in I_\mathbf{p}$. Indeed*

i) if $\pi_2(G/K)=\mathbf{Z}$ then $[\varphi_\alpha]=\pm1$ if α is long and ±2 otherwise;

ii) if $\pi_2(G/K)=\mathbf{Z}_2$ then $[\varphi_\alpha]=1$ if α is long.

Remarks. i) If $G\neq\mathbf{G}_2$ and $\pi_2(G/K)=\mathbf{Z}_2$ then $[\varphi_\alpha]=0$ for α short. However, all root spheres in $\mathbf{G}_2/\mathbf{SO}(4)$ are homotopically non-trivial since the ratio of the squared root lengths is three.

ii) We note that the homotopy class of φ_α depends only on the length of α.

iii) In chapter 7, we shall show that if $\alpha\in I_\mathbf{p}$ is long then φ_α is a stable harmonic map.

F. Tables

It is clear that the results of section D provide an effective method of listing the simply connected type I symmetric spaces together with their second homotopy groups. As a convenience to the Reader, we present such a list below in the form of two tables; one detailing the inner symmetric spaces and the other detailing the non-inner ones.

Inner symmetric spaces of type I			
Type	$\pi_2=\mathbf{Z}$	$\pi_2=0$	$\pi_2=\mathbf{Z}_2$
$\mathbf{A}_{n-1}$	$\frac{\mathbf{SU}(n)}{\mathbf{S}(\mathbf{U}(r)\times\mathbf{U}(n-r))}$ $1\le r<n$		
$\mathbf{B}_n$	$\frac{\mathbf{SO}(2n+1)}{\mathbf{SO}(2)\times\mathbf{SO}(2n-1)}$	S^{2n}	$\frac{\mathbf{SO}(2n+1)}{\mathbf{SO}(k)\times\mathbf{SO}(2n+1-k)}$ $2<k\le n$
$\mathbf{C}_n$	$\frac{\mathbf{Sp}(n)}{\mathbf{U}(n)}$	$\frac{\mathbf{Sp}(n)}{\mathbf{Sp}(k)\times\mathbf{Sp}(n-k)}$	
$\mathbf{D}_n$	$\frac{\mathbf{SO}(2n)}{\mathbf{U}(n)}$ $\frac{\mathbf{SO}(2n)}{\mathbf{SO}(2)\times\mathbf{SO}(2(n-1))}$		$\frac{\mathbf{SO}(2n)}{\mathbf{SO}(2k)\times\mathbf{SO}(2(n-k))}$ $2<2k\le n$
$\mathbf{E}_6$	$\frac{\mathbf{E}_6}{\mathbf{Spin}(10)\mathbf{U}(1)}$		$\frac{\mathbf{E}_6}{\mathbf{SU}(6)\mathbf{Sp}(1)}$
$\mathbf{E}_7$	$\frac{\mathbf{E}_7}{\mathbf{E}_6\mathbf{U}(1)}$		$\frac{\mathbf{E}_7}{\mathbf{Spin}(12)\mathbf{Sp}(1)}$ $\frac{\mathbf{E}_7}{\mathbf{SU}(8)/\mathbf{Z}_2}$
$\mathbf{E}_8$			$\frac{\mathbf{E}_8}{\mathbf{E}_7\mathbf{Sp}(1)}$ $\frac{\mathbf{E}_8}{\mathbf{Spin}(16)/\mathbf{Z}_2}$
$\mathbf{F}_4$		$\frac{\mathbf{F}_4}{\mathbf{Spin}(9)}$	$\frac{\mathbf{F}_4}{\mathbf{Sp}(3)\mathbf{Sp}(1)}$
$\mathbf{G}_2$			$\frac{\mathbf{G}_2}{\mathbf{SO}(4)}$

Non-inner symmetric spaces of type I		
Type	$\pi_2=0$	$\pi_2=\mathbf{Z}_2$
$\mathbf{A}_{n-1}$	$\frac{\mathbf{SU}(2k)}{\mathbf{Sp}(k)}$ $2k=n$	$\frac{\mathbf{SU}(n)}{\mathbf{SO}(n)}$
$\mathbf{D}_n$	S^{2n-1}	$\frac{\mathbf{SO}(2n)}{\mathbf{SO}(k)\times\mathbf{SO}(2n-k)}$ k odd, $1<k\le n$
$\mathbf{E}_6$	$\frac{\mathbf{E}_6}{\mathbf{F}_4}$	$\frac{\mathbf{E}_6}{\mathbf{Sp}(4)/\mathbf{Z}_2}$

Chapter 4. Flag Manifolds

Another important class of homogeneous spaces is that of *flag manifolds*. These are the manifolds of the form G/H where G is a compact Lie group and H is the centraliser of a torus of G. In case that G is semi-simple, these spaces exhaust the compact Kählerian G-spaces [6].

We shall see that flag manifolds are intimately related to the twistor theory of symmetric spaces. In particular, each flag manifold is fibred in a canonical way over an inner Riemannian symmetric space to provide a natural class of twistor fibrations which we shall treat in the sequel.

Throughout this chapter, G will denote a compact semi-simple Lie group. The extension to the general (i.e reductive) case is left to the Reader.

A. Parabolic subalgebras

Flag manifolds have another realisation as homogeneous spaces since the complexification of G acts transitively on G-flag manifolds with parabolic subgroups as stabilizers. It is from the interplay between these real and complex realisations that the rich geometry of such manifolds derives.

Therefore we begin by studying the infinitesimal situation and investigate the properties of parabolic subalgebras and their interaction with symmetric decompositions.

Preliminaries

Let $\mathbf{g}^{\mathbf{C}}$ be a complex semi-simple Lie algebra with Killing form denoted by B. Given a subspace $V \subset \mathbf{g}^{\mathbf{C}}$, we shall denote by $V^{\perp}$ the polar of V with respect to B.

Definition.
(a) A subalgebra $\mathbf{b}$ of $\mathbf{g}^{\mathbf{C}}$ is a *Borel subalgebra* if it is a maximal solvable subalgebra of $\mathbf{g}^{\mathbf{C}}$.
(b) A subalgebra $\mathbf{q}$ of $\mathbf{g}^{\mathbf{C}}$ is a *parabolic subalgebra* if it contains a Borel subalgebra.

The relationship between such subalgebras and root systems is given in the following theorem (c.f. [46]).

Theorem 4.1 *Let* $\mathbf{a}$ *be a CSA for* $\mathbf{g}^{\mathbf{C}}$ *and* Δ^+ *a positive root system with simple roots* $\alpha_1,\ldots,\alpha_l$. *Then*

(a)

$$\mathbf{b} = \mathbf{a} \oplus \sum_{\alpha\in\Delta^+} \mathbf{g}^{\alpha}$$

is a Borel subalgebra and any other Borel subalgebra is conjugate to $\mathbf{b}$*;*

(b) each subset I *of* $\{1,\ldots,l\}$ *defines a 'height' function* n_I *on* Δ *by*

$$n_I(\alpha) = \sum_{i\in I} n_i$$

for $\alpha = \sum_{i=1}^{l} n_i \alpha_i$ *and then*

$$\mathbf{q}_I = \mathbf{a} \oplus \sum_{n_I(\alpha)\geq 0} \mathbf{g}^\alpha$$

is a parabolic subalgebra. Moreover, every parabolic subalgebra is conjugate to a $\mathbf{q}_I$ *for a unique subset* I *of* $\{1,\ldots,l\}$.

There are thus a finite number, 2^l, of conjugacy classes of parabolic subalgebras ($\mathbf{g}^{\mathbf{C}}$ itself is regarded as the parabolic subalgebra corresponding with the empty set, while the Borel subalgebra $\mathbf{b}$ corresponds with $I=\{1,\ldots,l\}$). Parabolic subalgebras obviously contain Cartan subalgebras of $\mathbf{g}^{\mathbf{C}}$ and, moreover, are easily shown to be of the form $\mathbf{q}_I$ with respect to a suitably chosen Weyl chamber for any CSA they contain.
We observe that the nilradical (i.e maximal nilpotent ideal) of $\mathbf{q}_I$ is given by

$$\sum_{n_I(\alpha)>0} \mathbf{g}^\alpha .$$

A useful criterion for detecting parabolic subalgebras is given by lemma (4.2) in Grothendieck's paper [40]:

Proposition 4.2. *Let* $\mathbf{n}\subset\mathbf{g}^{\mathbf{C}}$ *be a nilpotent subalgebra such that* $\mathbf{n}^\perp$ *is also a subalgebra. Then* $\mathbf{n}^\perp$ *is a parabolic subalgebra with nilradical* $\mathbf{n}$.

Thus parabolic subalgebras are subalgebras whose polar is a nilpotent subalgebra. In fact the polar can be taken with respect to any invariant non-degenerate bilinear symmetric form on $\mathbf{g}^{\mathbf{C}}$.
If $\mathbf{q}$ is a parabolic subalgebra with nilradical $\mathbf{n}$, then the quotient algebra $\mathbf{q}/\mathbf{n}$ is reductive.

Definition. A subalgebra $\mathbf{h}$ of a parabolic subalgebra $\mathbf{q}$ for which we have a direct sum decomposition

$$\mathbf{q} = \mathbf{h}\oplus\mathbf{n}$$

is called a *reductive factor* for $\mathbf{q}$.

Reductive factors are isomorphic to $\mathbf{q}/\mathbf{n}$ and so are reductive Lie algebras. We may obtain a reductive factor for $\mathbf{q}_I$ by taking

$$\mathbf{h} = \mathbf{a}\oplus \sum_{n_I(\alpha)=0} \mathbf{g}^\alpha .$$

We now show that, if $\mathbf{g}^{\mathbf{C}}$ is simple, a reductive factor acts irreducibly on the centre of the nilradical. This fact (for which we could not find a proof in the literature) has applications to the stability of harmonic 2-spheres in symmetric spaces as we shall see in Chapter 7.

Proposition 4.3. *Let* $\mathbf{g}^{\mathbf{C}}$ *be a simple Lie algebra and* $\mathbf{q}$ *a parabolic subalgebra with nilradical* $\mathbf{n}$ *and reductive factor* $\mathbf{h}$. *Then* $\mathrm{ad}\mathbf{h}$ *preserves the centre,* $\mathbf{z}$, *of* $\mathbf{n}$ *and the action of* $\mathbf{h}$ *on* $\mathbf{z}$ *is irreducible.*

Proof. Since $\mathbf{n}$ is an ideal of $\mathbf{q}$ we have $[\mathbf{q},\mathbf{z}]\subset\mathbf{z}$ and in particular $[\mathbf{h},\mathbf{z}]\subset\mathbf{z}$. Thus $\mathbf{z}$ is a $\mathbf{q}$-module. Further, since $\mathbf{n}$ acts trivially on $\mathbf{z}$ by definition and $\mathbf{h}$ acts as $\mathbf{q}/\mathbf{n}$, it will suffice to show that $\mathbf{z}$ is an irreducible $\mathbf{q}$-module.
For this, we begin by picking a CSA $\mathbf{a}\subset\mathbf{q}$ and a positive root system Δ^+ so that $\mathbf{q}=\mathbf{q}_I$ for some subset I. The adjoint representation of a simple Lie algebra is irreducible and so $\mathrm{ad}\mathbf{g}^{\mathbf{C}}$ has a

unique highest weight which is the (lexicographically) highest root θ. Clearly, $[\mathbf{n}, \mathbf{g}^\theta] = 0$ so that $\mathbf{g}^\theta \subset \mathbf{z}$ whence the (necessarily irreducible) $\mathbf{q}$-module $\mathbf{z}'$ generated by $\mathbf{g}^\theta$ is contained in $\mathbf{z}$. It remains to show that $\mathbf{z} \subset \mathbf{z}'$.

Now, any irreducible $\mathbf{g}^{\mathbf{C}}$-module is generated from the highest weight space by the action of vectors in the root spaces of the negatives of the simple roots. Thus, for any root α,

$$\mathbf{g}^\alpha = [\mathbf{g}^{-\alpha_{i_1}}, [\ldots, [\mathbf{g}^{-\alpha_{i_k}}, \mathbf{g}^\theta]\ldots]]$$

for suitable $i_1, \ldots, i_k$. It follows that every intermediate bracket is a root space and so theorem (3.4) allows us to invert this formula to give

$$\mathbf{g}^\theta = [\mathbf{g}^{\alpha_{i_k}}, [\ldots, [\mathbf{g}^{\alpha_{i_1}}, \mathbf{g}^\alpha]\ldots]] \,. \tag{1}$$

$\mathbf{z}$ is a sum of root spaces so let $\mathbf{g}^\alpha \subset \mathbf{z}$. From (1) we see that no i_j is in I since otherwise the right-hand side would vanish. Hence each $\mathbf{g}^{-\alpha_{i_j}} \subset \mathbf{q}$ and so $\mathbf{g}^\alpha \subset \mathbf{z}'$ and the theorem follows. □

Remark. In [79] a similar result due to Kostant is proved.

Compact real forms and canonical elements

Let us now consider the additional structure given by a compact real form $\mathbf{g}$ of $\mathbf{g}^{\mathbf{C}}$. We denote by $\xi \mapsto \bar{\xi}$ the complex conjugation on $\mathbf{g}^{\mathbf{C}}$ with respect to the real form $\mathbf{g}$.

If $\mathbf{q} \subset \mathbf{g}^{\mathbf{C}}$ is a parabolic subalgebra then $\mathbf{g} \cap \mathbf{q}$ contains a maximal torus $\mathbf{t}$ of $\mathbf{g}$ and $\mathbf{t}^{\mathbf{C}}$ is a CSA of $\mathbf{g}^{\mathbf{C}}$ contained in $\mathbf{q}$.

We recall from Chapter 3 that $\Delta(\mathbf{g}^{\mathbf{C}}, \mathbf{t}^{\mathbf{C}}) \subset \sqrt{-1}\mathbf{t}^*$ so that for each root α we have

$$\overline{\mathbf{g}^\alpha} = \mathbf{g}^{-\alpha} \,.$$

Since $\mathbf{q}$ must have the form $\mathbf{q}_I$ for some subset I of simple roots with respect to $\mathbf{t}^{\mathbf{C}}$, we see that a reductive factor for $\mathbf{q}$ is given by

$$\mathbf{q} \cap \bar{\mathbf{q}} = \mathbf{t}^{\mathbf{C}} \oplus \sum_{n_I(\alpha)=0} \mathbf{g}^\alpha \,.$$

So putting $\mathbf{h} = \mathbf{q} \cap \bar{\mathbf{q}} \cap \mathbf{g}$, we have

$$\mathbf{g}^{\mathbf{C}} = \mathbf{h}^{\mathbf{C}} \oplus \mathbf{n} \oplus \bar{\mathbf{n}} \,, \qquad \mathbf{q} = \mathbf{h}^{\mathbf{C}} \oplus \mathbf{n} \,,$$

where $\mathbf{n}$ is the nilradical of $\mathbf{q}$. Clearly, $\mathbf{h}$ is the centraliser of the torus

$$\{X \in \mathbf{t} : \alpha_i(X) = 0 \quad \text{for all } i \notin I\} \,.$$

Conversely, if $\mathbf{h}$ is the centraliser of some torus in $\mathbf{g}$, it is straight-forward to show that $\mathbf{h} = \mathbf{g} \cap \mathbf{q}$ for at least one parabolic subalgebra $\mathbf{q}$ of $\mathbf{g}^{\mathbf{C}}$.

Given a parabolic subalgebra $\mathbf{q}$, we distinguish a certain element of the centre of $\mathbf{h}$, uniquely determined by $\mathbf{q}$, whose adjoint action completely determines $\mathbf{q}$.

Theorem 4.4. *Let $\mathbf{q} \subset \mathbf{g}^{\mathbf{C}}$ be a parabolic subalgebra with $\mathbf{h} \subset \mathbf{g}$, $\mathbf{n}$ as above. Then there is a unique element ξ of the centre of $\mathbf{h}$ with the following properties:*

(i) the eigenvalues of $\mathrm{ad}\xi$ lie in $\sqrt{-1}\mathbf{Z}$;

(ii) for $r \in \mathbf{Z}$, let $\mathbf{g}_r$ denote the $\sqrt{-1}r$-eigenspace of $\mathrm{ad}\xi$. Further, let $\mathbf{n}^{(r)}$ denote the r-th step of the central descending series of $\mathbf{n}$ i.e. $\mathbf{n}^{(1)} = \mathbf{n}$, $\mathbf{n}^{(r+1)} = [\mathbf{n}^{(r)}, \mathbf{n}]$ for $r \geq 1$. Then

$$\mathbf{q} = \sum_{i \geq 0} \mathbf{g}_i \,, \qquad \mathbf{n}^{(r)} = \sum_{i \geq r} \mathbf{g}_i \,. \tag{2}$$

We call ξ the *canonical element* of $\mathbf{q}$.

Proof. That ξ is unique is clear, since (2) completely determines $\mathrm{ad}\xi$ and $\mathbf{g}$, being semi-simple, has no centre.

To show existence, fix a maximal torus $\mathbf{t}$ in $\mathbf{h}$ together with a positive root system so that $\mathbf{q}=\mathbf{q}_I$. Now we define dual vectors $\{\xi_1,\dots,\xi_l\}\subset\mathbf{t}$ to the simple roots $\{\alpha_1,\dots,\alpha_l\}\subset\sqrt{-1}\mathbf{t}^*$ by

$$\alpha_i(\xi_j)=\sqrt{-1}\delta_{ij}$$

and set $\xi=\sum_{i\in I}\xi_i$. Thus $\mathrm{ad}\xi$ has eigenvalue $\sqrt{-1}n_I(\alpha)\in\sqrt{-1}\mathbf{Z}$ on $\mathbf{g}^\alpha$ and in particular, ξ centralises $\mathbf{h}$. We claim that ξ is our canonical element.

Clearly, we have

$$\mathbf{q}=\sum_{i\geq 0}\mathbf{g}_i\,,\qquad \mathbf{n}=\sum_{i\geq 1}\mathbf{g}_i$$

and since $[\mathbf{g}_i,\mathbf{g}_j]\subset\mathbf{g}_{i+j}$ we have

$$\mathbf{n}^{(r)}\subset\sum_{i\geq r}\mathbf{g}_i\ .$$

For the reverse inclusion it suffices to show that $\mathbf{g}_r\subset\mathbf{n}^{(r)}$. But if $\mathbf{g}^\alpha\subset\mathbf{g}_r$, we may write $\alpha=\alpha_{i_1}+\dots+\alpha_{i_k}$ with each partial sum a root and precisely r of the i_j in I whence

$$\mathbf{g}^\alpha=[\mathbf{g}^{\alpha_{i_k}},[\dots,[\mathbf{g}^{\alpha_{i_2}},\mathbf{g}^{\alpha_{i_1}}]\dots]]\ .$$

Now, for $i\in I$, $\mathbf{g}^{\alpha_i}\subset\mathbf{n}$ whence

$$[\mathbf{g}^{\alpha_i},\mathbf{n}^{(j)}]\subset\mathbf{n}^{(j+1)}\ ,$$

while, for $i\notin I$, $\mathbf{g}^{\alpha_i}\subset\mathbf{h}^{\mathbf{C}}$ so that

$$[\mathbf{g}^{\alpha_i},\mathbf{n}^{(j)}]\subset\mathbf{n}^{(j)}\ ,$$

from which we obtain $\mathbf{g}^\alpha\subset\mathbf{n}^{(r)}$ and the theorem is proved. □

Remarks. (i) We observe that the theorem implies that $\mathbf{g}_1$ generates $\mathbf{n}$ and ,indeed,

$$\mathbf{g}_r=[\mathbf{g}_1,[\dots,[\mathbf{g}_1,\mathbf{g}_1]\dots]] \tag{3}$$

with $\mathbf{g}_1$ appearing r times.

(ii) We note that for $i>1$,

$$\mathbf{g}_i=\mathbf{n}^{(i)}\cap\bar{\mathbf{n}}^{(i+1)\perp}\ . \tag{4}$$

(iii) For $\mathbf{q}=\mathbf{q}_I\subset\mathbf{g}^{\mathbf{C}}$, it is easy to show that the centre of $\mathbf{h}$ is $|I|$-dimensional (it is spanned by $\{\xi_i\colon i\in I\}$) while the Dynkin diagram of the semi-simple part of $\mathbf{h}^{\mathbf{C}}$ is obtained from that of $\mathbf{g}^{\mathbf{C}}$ by striking out the vertices corresponding to $\{\alpha_i\colon i\in I\}$.

Canonical symmetric decompositions and complex structures

We now show that the canonical element of a parabolic subalgebra induces a canonical symmetric decomposition on $\mathbf{g}$.

Let $\mathbf{q}\subset\mathbf{g}^{\mathbf{C}}$ be a parabolic subalgebra with $\mathbf{h}=\mathbf{q}\cap\mathbf{g}$ and canonical element ξ. Then, since $\sqrt{-1}\mathrm{ad}\xi$ has integer eigenvalues, we see that $\tau_\xi=\mathrm{Ad}\exp\pi\xi$ is an involutive automorphism of $\mathbf{g}$. Further, if $\mathbf{g}=\mathbf{k}\oplus\mathbf{p}$ is the corresponding symmetric decomposition, then

$$\mathbf{k}^{\mathbf{C}} = \sum_{r\in 2\mathbf{Z}} \mathbf{g}_r\ ,\quad \mathbf{p}^{\mathbf{C}} = \sum_{r\in 2\mathbf{Z}+1} \mathbf{g}_r\ ,$$

where, as above, $\mathbf{g}_r$ is the $\sqrt{-1}r$-eigenspace of $\mathrm{ad}\xi$.

We call τ_ξ the *canonical involution for* $\mathbf{q}$.

Let us settle the question of which involutive automorphisms arise in this way. By construction, τ_ξ is an *inner* involution and in fact this condition is sufficient.

Proposition 4.5. *An involution of* $\mathbf{g}$ *is canonical for some parabolic subalgebra of* $\mathbf{g}^{\mathbf{C}}$ *if and only if it is inner.*

Proof. Let τ be an inner involution of $\mathbf{g}$ with symmetric decomposition

$$\mathbf{g} = \mathbf{k}\oplus\mathbf{p}\ .$$

Since τ is inner, a maximal torus of $\mathbf{k}$ is maximal in $\mathbf{g}$. Let $\mathbf{t}$ be such a torus and let $\Delta^+\subset\Delta(\mathbf{g}^{\mathbf{C}},\mathbf{t}^{\mathbf{C}})$ be a positive root system with simple roots $\alpha_1,\ldots,\alpha_l$. Define $I\subset\{1,\ldots,l\}$ by

$$I = \{i\colon \mathbf{g}^{\alpha_i}\subset\mathbf{p}\}\ .$$

If $\mathbf{p}\neq\{0\}$, I is nonempty. We claim that τ is canonical for $\mathbf{q}_I$.

For this, observe that if $n_I(\alpha)=r$ we may write

$$\mathbf{g}^\alpha = [\mathbf{g}^{\alpha_{i_1}}, [\ldots, [\mathbf{g}^{\alpha_{i_{k-1}}}, \mathbf{g}^{\alpha_{i_k}}]\ldots]]$$

with precisely r of the i_j in I. Thus, precisely r of the $\mathbf{g}^{\alpha_{i_j}}$ are contained in $\mathbf{p}$ so that τ has eigenvalue $(-1)^r$ on $\mathbf{g}^\alpha$. So $\mathbf{g}^\alpha\subset\mathbf{p}$ if and only if $n_I(\alpha)$ is odd and the claim follows immediately. □

Thus we see that a parabolic subalgebra of $\mathbf{g}^{\mathbf{C}}$ provides an inner involution on the compact real form $\mathbf{g}$ and conversely. In fact, more is true: a parabolic subalgebra $\mathbf{q}$ provides a complex structure on the corresponding $\mathbf{p}$ with $\sqrt{-1}$-eigenspace $\mathbf{p}^+$ given by

$$\mathbf{p}^+ = \sum_{r\geq 0}\mathbf{g}_{2r+1} = \mathbf{p}^{\mathbf{C}}\cap\mathbf{q}\ .$$

Further, we observe that $[\mathbf{p}^+,\mathbf{p}^+]$ is isotropic and from remark (i) following theorem (4.4) we see that the nilradical of $\mathbf{q}$ is given by

$$[\mathbf{p}^+,\mathbf{p}^+]\oplus\mathbf{p}^+\ .$$

These properties of $\mathbf{p}^+$ are enough to characterise $\mathbf{q}$ and a powerful converse is available.

Indeed, let τ be an involution with symmetric decomposition $\mathbf{g}=\mathbf{k}\oplus\mathbf{p}$. The proofs of the next two results is deferred until Chapter 5 where stronger results will be proved. (See lemma (5.1) and proposition (5.2).)

Lemma 4.6. *Let* $\mathbf{p}^+$ *be a maximally isotropic subspace of* $\mathbf{p}^{\mathbf{C}}$.

(i) $[\mathbf{p}^+,\mathbf{p}^+]\subset\mathbf{k}^{\mathbf{C}}$ *is isotropic if and only if* $\mathbf{p}^+$ *satisfies*

$$[[\mathbf{p}^+,\mathbf{p}^+],\mathbf{p}^+]\subset\mathbf{p}^+. \tag{5}$$

(ii) Let $\mathbf{p}^+$ *satisfy (5) and put* $\mathbf{h}=\{\xi\in\mathbf{k}\colon[\xi,\mathbf{p}^+]\subset\mathbf{p}^+\}$. *Then if* $\mathbf{m}=\mathbf{h}^\perp\cap\mathbf{k}$ *we have*

$$\mathbf{m}^{\mathbf{C}} = [\mathbf{p}^+,\mathbf{p}^+]\oplus\overline{[\mathbf{p}^+,\mathbf{p}^+]}.$$

In particular, $[\mathbf{p}^+,\mathbf{p}^+]$ *is a maximal isotropic subspace of* $\mathbf{m}^{\mathbf{C}}$ *and the polar of* $[\mathbf{p}^+,\mathbf{p}^+]$ *in* $\mathbf{k}^{\mathbf{C}}$ *is*

$\mathbf{h}^{\mathbf{C}} \oplus [\mathbf{p}^+, \mathbf{p}^+]$.

Proposition 4.7. *Let* $\mathbf{p}^+$ *satisfy (5) and set* $\mathbf{h} = \{\xi \in \mathbf{k} : [\xi, \mathbf{p}^+] \subset \mathbf{p}^+\}$. *Then* $\mathbf{h}$ *is the centraliser of a torus in* $\mathbf{k}$.

We can now show that if τ is inner then a $\mathbf{p}^+$ satisfying (5) generates a parabolic subalgebra for which τ is canonical.

Theorem 4.8. *Let* τ *be an inner involution on* $\mathbf{g}$ *with symmetric decomposition* $\mathbf{g} = \mathbf{k} \oplus \mathbf{p}$. *Let* $\mathbf{p}^+ \subset \mathbf{p}^{\mathbf{C}}$ *be a maximally isotropic subspace of* $\mathbf{p}^{\mathbf{C}}$ *such that* $[\mathbf{p}^+, \mathbf{p}^+]$ *is isotropic. Then there is a unique* τ*-stable parabolic subalgebra* $\mathbf{q}$ *of* $\mathbf{g}^{\mathbf{C}}$ *with the following properties:*

(i) τ *is canonical for* $\mathbf{q}$*;*

(ii) $\mathbf{p}^+ = \mathbf{p}^{\mathbf{C}} \cap \mathbf{q}$*;*

(iii) the nilradical of $\mathbf{q}$ *is given by*

$$[\mathbf{p}^+, \mathbf{p}^+] \oplus \mathbf{p}^+ .$$

Proof. Denote $[\mathbf{p}^+, \mathbf{p}^+] \oplus \mathbf{p}^+$ by $\mathbf{n}$ and $\mathbf{n}^\perp$ by $\mathbf{q}$.

We claim that $\mathbf{q}$ is a parabolic subalgebra. Given this, it is immediate that

(a) $\mathbf{q}$ is τ-stable since $\mathbf{n}$ is;

(b) $\mathbf{p}^+ = \mathbf{p}^{\mathbf{C}} \cap \mathbf{q}$, since $\mathbf{p}^+$ is maximally isotropic in $\mathbf{p}^{\mathbf{C}}$;

(c) the nilradical of $\mathbf{q}$ is $\mathbf{q}^\perp = [\mathbf{p}^+, \mathbf{p}^+] \oplus \mathbf{p}^+$.

To prove the claim, we first observe that $\mathbf{n}$ is a subalgebra since, from lemma (4.6i), we have

$$[[\mathbf{p}^+, \mathbf{p}^+], \mathbf{p}^+] \subset \mathbf{p}^+$$

and hence that

$$[[\mathbf{p}^+, \mathbf{p}^+], [\mathbf{p}^+, \mathbf{p}^+]] \subset [\mathbf{p}^+, \mathbf{p}^+] .$$

Again, from lemma (4.6ii), putting $\mathbf{h} = \{\eta \in \mathbf{k} : [\eta, \mathbf{p}^+] \subset \mathbf{p}^+\}$, we see that

$$\mathbf{q} = \mathbf{h} \oplus \mathbf{n}$$

so that $\mathbf{q}$ is also a subalgebra.

Further, from proposition (4.7), we see that $\mathbf{h}$ is the centraliser of a torus in $\mathbf{k}$ and so contains a maximal torus of $\mathbf{k}$ which is in fact maximal in $\mathbf{g}$ since τ is inner. Let $\mathbf{t}$ be such a torus and define $\Phi \subset \Delta(\mathbf{g}^{\mathbf{C}}, \mathbf{t}^{\mathbf{C}})$ by

$$\Phi = \{\alpha \in \Delta : \mathbf{g}^\alpha \subset \mathbf{n}\} .$$

Since $\mathbf{n}$ is isotropic, we have $\mathbf{n} \cap \bar{\mathbf{n}} = \{0\}$ so that $\Phi \cap -\Phi = \varnothing$. Also, Φ is closed since $\mathbf{n}$ is a subalgebra so we may conclude that Φ is contained in some positive root system. From this it is clear that $\mathbf{n}$ is nilpotent and thus, by proposition (4.2), $\mathbf{q}$ is parabolic. This proves the claim.

It remains to show that τ is canonical for $\mathbf{q}$. Since

$$\mathbf{q} \cap \bar{\mathbf{q}} = \mathbf{h}^{\mathbf{C}} \subset \mathbf{k}^{\mathbf{C}} ,$$

the canonical element ξ of $\mathbf{q}$ is contained in $\mathbf{k}$. Thus the eigenspaces of $\mathrm{ad}\xi$ are τ-stable. Further,

$$\mathbf{n} \cap \mathbf{k}^{\mathbf{C}} = [\mathbf{p}^+, \mathbf{p}^+] \subset [\mathbf{n}, \mathbf{n}]$$

whence, by theorem (4.4ii), we have $\mathbf{g}_1 \subset \mathbf{p}^+$. Thus, by (3), $\mathbf{g}_r$ is contained in $\mathbf{p}^{\mathbf{C}}$ if and only if r is odd and so τ is the canonical involution for $\mathbf{q}$. □

Remark. The interaction between non-inner involutions and parabolic subalgebras will be considered in an appendix to this chapter.

The non-compact dual

Let us now consider the interaction of parabolic subalgebras with *non-compact* real forms of $\mathbf{g}^{\mathbf{C}}$. Let $\mathbf{a} \subset \mathbf{g}_R$ be a Cartan subalgebra of $\mathbf{g}_R$ (thus $\mathbf{a}^{\mathbf{C}}$ is a CSA for $\mathbf{g}_R^{\mathbf{C}}$). We say that $\mathbf{a}$ is *compact* if $\mathbf{a}$ is point-wise fixed by some Cartan involution of $\mathbf{g}_R$. For this it is necessary and sufficient that B be negative definite on $\mathbf{a}$.

Lemma 4.9. *A compact CSA is fixed by precisely one Cartan involution.*

Proof. If $\mathbf{g}_R = \mathbf{k} \oplus \mathbf{p}$ is a Cartan decomposition with $\mathbf{a} \subset \mathbf{k}$, then both $\mathbf{k}^{\mathbf{C}}$ and $\mathbf{p}^{\mathbf{C}}$ are ad$\mathbf{a}^{\mathbf{C}}$-stable and so for $\alpha \in \Delta(\mathbf{g}_R^{\mathbf{C}}, \mathbf{a}^{\mathbf{C}})$, we have either $\mathbf{g}^\alpha \subset \mathbf{k}^{\mathbf{C}}$ or $\mathbf{g}^\alpha \subset \mathbf{p}^{\mathbf{C}}$. Further, if $X \in \mathbf{g}^\alpha$, then $B(X, \bar{X})$ is strictly positive for $\mathbf{g}^\alpha \subset \mathbf{p}^{\mathbf{C}}$ and strictly negative for $\mathbf{g}^\alpha \subset \mathbf{k}^{\mathbf{C}}$. Thus

$$\mathbf{p}^{\mathbf{C}} = \sum \{\mathbf{g}^\alpha : B(X, \bar{X}) > 0 \text{ for } X \in \mathbf{g}^\alpha\}$$

$$\mathbf{k}^{\mathbf{C}} = \mathbf{a} \oplus \sum \{\mathbf{g}^\alpha : B(X, \bar{X}) < 0 \text{ for } X \in \mathbf{g}^\alpha\}$$

and so the Cartan decomposition is uniquely determined. □

Remarks. (i) A compact CSA of $\mathbf{g}_R$ is a maximal torus in the compact dual of $\mathbf{g}_R$

(ii) Of course, $\mathbf{g}_R$ need not admit a compact CSA. Indeed, $\mathbf{g}_R$ will do so if and only if its Cartan involution is inner ([42, p. 424]).

Now let $\mathbf{q} \subset \mathbf{g}^{\mathbf{C}}$ be a parabolic subalgebra and $\mathbf{g}_R$ be a non-compact real form of $\mathbf{g}^{\mathbf{C}}$. Then $\mathbf{q} \cap \mathbf{g}_R$ contains a CSA of $\mathbf{g}_R$ [80].

Definition. The pair $(\mathbf{g}_R, \mathbf{q})$ is a *compact pair* if $\mathbf{q} \cap \mathbf{g}_R$ contains a compact CSA.

We can see that for a compact pair, we may construct a canonical element as above. Indeed, if $\mathbf{a} \subset \mathbf{q} \cap \mathbf{g}_R$ is a compact CSA and τ is the Cartan involution fixing $\mathbf{a}$, then, for each $X \in \mathbf{a}$, AdX is skew with respect to the inner product B_τ and so has imaginary eigenvalues. We may now repeat the analysis of theorem (4.4) to get:

Theorem 4.10.. *Let* $\mathbf{q} \subset \mathbf{g}^{\mathbf{C}}$ *be a parabolic subalgebra with nilradical* $\mathbf{n}$ *and* $\mathbf{g}_R$ *a non-compact real form of* $\mathbf{g}^{\mathbf{C}}$ *so that* $(\mathbf{g}_R, \mathbf{q})$ *is a compact pair. Put* $\mathbf{h} = \mathbf{q} \cap \mathbf{g}_R$. *Then* $\mathbf{h}^{\mathbf{C}}$ *is a reductive factor for* $\mathbf{q}$ *and there is a unique element* ξ *of the centre of* $\mathbf{h}$ *with the following properties:*

(i) the eigenvalues of adξ *lie in* $\sqrt{-1}\mathbf{Z}$;

(ii) for $r \in \mathbf{Z}$, *let* $\mathbf{g}_r$ *denote the* $\sqrt{-1}r$-*eigenspace of* adξ. *Further, let* $\mathbf{n}^{(r)}$ *denote the r-th step of the central descending series of* $\mathbf{n}$ *i.e.* $\mathbf{n}^{(1)} = \mathbf{n}$, $\mathbf{n}^{(r+1)} = [\mathbf{n}^{(r)}, \mathbf{n}]$ *for* $r \geq 1$. *Then*

$$\mathbf{q} = \sum_{i \geq 0} \mathbf{g}_i, \qquad \mathbf{n}^{(r)} = \sum_{i \geq r} \mathbf{g}_i.$$

We call ξ the *canonical element* of $(\mathbf{g}_R, \mathbf{q})$.

Now, setting $\tau_\xi = \mathrm{Ad}\exp\pi\xi$ provides an involution on $\mathbf{g}_R$ which we call the *canonical involution* for $(\mathbf{g}_R, \mathbf{q})$.

Definition. A compact pair is a *canonical pair* if the canonical involution is a Cartan involution.

Remark. We can see from formula (3) that a compact pair is canonical if the Killing form is negative definite on $\mathbf{h}$ and positive definite on $\mathbf{g}_R \cap (\mathbf{g}_1 \oplus \mathbf{g}_{-1})$.

The canonical involution on a canonical pair $(\mathbf{g}_R, \mathbf{q})$ provides a compact dual to $\mathbf{g}_R$ and we observe that the canonical element of $(\mathbf{g}_R, \mathbf{q})$ coincides with the canonical element of $\mathbf{q}$ with respect to this compact real form. From this we easily obtain the following corollary to proposition (4.5):

Proposition 4.11. *There is a parabolic subalgebra* $\mathbf{q}$ *for which* $(\mathbf{g}_R, \mathbf{q})$ *is a canonical pair if and only if the Cartan involution of* $\mathbf{g}_R$ *is inner.*

Similarly, we apply theorem (4.8) to the compact dual to obtain:

Theorem 4.12 *Let* $\mathbf{g}_R$ *be a non-compact semi-simple real Lie algebra with inner Cartan involution* τ *and Cartan decomposition* $\mathbf{g}_R = \mathbf{k} \oplus \mathbf{p}$. *Let* $\mathbf{p}^+ \subset \mathbf{p}^C$ *be a maximally isotropic subspace of* $\mathbf{p}^C$ *such that* $[\mathbf{p}^+, \mathbf{p}^+]$ *is isotropic. Then there is a unique* τ*-stable parabolic subalgebra* $\mathbf{q}$ *of* $\mathbf{g}^C$ *with the following properties:*

(i) $(\mathbf{g}_R, \mathbf{q})$ *is a canonical pair for which* τ *is the canonical involution.*

(ii) $\mathbf{p}^+ = \mathbf{p}^C \cap \mathbf{q}$;

(iii) the nilradical of $\mathbf{q}$ *is given by*

$$[\mathbf{p}^+, \mathbf{p}^+] \oplus \mathbf{p}^+ \ .$$

Interpretation. Let N be a symmetric space of semi-simple type with involution τ. Suppose that $j \in J(N)$ has (1,0)-space $\mathbf{p}^+$. From (2.14) and formula (3) of chapter 2, we see that the Nijenhuis tensor of J_1 vanishes at j if and only if $\mathbf{p}^+$ satisfies (5). Theorems (4.8) and (4.12) now show that each such j corresponds to a unique parabolic subalgebra. This will form the basis of our study of the zero set of the Nijenhuis tensor in the following chapter.

B. Flag manifolds and flag domains

We now introduce the global analogues of the concepts developed above.

Flag manifolds

Definition. Let G^C be a connected semi-simple complex Lie group. A *parabolic subgroup* of G^C is a complex Lie subgroup which is the normaliser of a parabolic subalgebra of $\mathbf{g}^C$.
A *flag manifold* is a homogeneous space of the form G^C/P with P a parabolic subgroup.

The following facts are well-known (c.f. [80]): all parabolic subgroups are connected and a subgroup is parabolic if and only if its Lie algebra is. Further, a parabolic subgroup is the normaliser in $G^{\mathbf{C}}$ of its Lie algebra. Thus, if $\mathbf{q}$ is the Lie algebra of a parabolic subgroup P, then we may realise the flag manifold $G^{\mathbf{C}}/P$ as the conjugacy class of $\mathbf{q}$.

If G is a compact real form of $G^{\mathbf{C}}$, then G acts transitively on $G^{\mathbf{C}}/P$ so that a flag manifold is diffeomorphic to the real coset space $G/G\cap P$. Further, $G\cap P$ is connected and the centraliser of a torus, while, conversely, if H is the centraliser of a torus in G, then $H=G\cap P$ for at least one parabolic subgroup P of $G^{\mathbf{C}}$.

A flag manifold is a homogeneous space with a considerable amount of extra structure which is determined by (the canonical element of) the corresponding parabolic subgroup. So let $F=G^{\mathbf{C}}/P=G/H$ be a flag manifold and ξ the canonical element of (the Lie algebra of) P. Then the Lie algebra of H is given by $\mathbf{h}=\mathbf{g}_0\cap\mathbf{g}$ and setting

$$\mathbf{m} = \sum_{r\neq 0}\mathbf{g}_r\cap\mathbf{g}$$

provides an $\mathrm{Ad}(H)$-invariant splitting for $\mathbf{g}$ and makes F into a reductive G-space.

Further, F has a G-invariant complex structure defined by

$$T^{1,0}F = \sum_{r>0}[\mathbf{g}_r] = [\mathbf{n}]$$

where $\mathbf{n}$ is the nilradical of the Lie algebra of P. This complex structure, which makes F anti-biholomorphic with the complex coset space $G^{\mathbf{C}}/P$, we denote by J_1.

Now let us define a map $\Xi: F\to\mathbf{g}$ by setting $\Xi(x)$ equal to the canonical element of the (parabolic) stabilizer of $x\in F$. Ξ is then a G-equivariant diffeomorphism of F onto the adjoint orbit of ξ in $\mathbf{g}$. Equivalently we may view Ξ as a section of the isotropy bundle $[\mathbf{h}]$ of F and so we call Ξ the *canonical section.*

Using Ξ, we may define a Kähler metric for J_1. Indeed, if β^F is the Maurer-Cartan form for F then

$$\langle X,Y\rangle = (\beta^F(X), |\mathrm{ad}\Xi|\,\beta^F(Y))$$

defines a G-invariant Hermitian metric on F with Kähler form given by

$$-\tfrac{1}{2}(\Xi,[\beta^F\wedge\beta^F])$$

which is easily seen to be closed.

Remark. Identifying $\mathbf{g}$ with $\mathbf{g}^*$ via the Killing form, Ξ is a diffeomorphism of F onto the co-adjoint orbit of ξ and our Kähler form is just the pull-back by Ξ of the Kostant-Kirillov-Souriau symplectic form.

Flag domains

Now let G_R be a non-compact real form of $G^{\mathbf{C}}$ with Lie algebra $\mathbf{g}_R$ and suppose that $\mathbf{g}_R$ has inner Cartan involution. Let F be a flag manifold which we view as a conjugacy class of parabolic subalgebras of $\mathbf{g}^{\mathbf{C}}$. In contrast to the compact case, G_R does not act transitively on F.

Definition. A *flag domain* is a G_R-orbit in a flag manifold F which is open in F.

Warning. Our definition of flag domain is rather more general than others in the literature. For instance, Wells-Wolf [78] require in addition that the stabilisers of points of a flag domain be compact.

An extensive study of real orbits in flag manifolds has been made by Wolf [80] and specialising theorems (4.5) and (5.4) of [80] to the case of inner Cartan involution we obtain:

Theorem 4.13. *Let G_R be a non-compact real form of $G^{\mathbf{C}}$ with inner Cartan involution and let F be a flag manifold. Then the G_R-orbit of $\mathbf{q}\in F$ is open if and only if $(\mathbf{g}_R,\mathbf{q})$ is a compact pair. Further, in this case, the stabiliser of $\mathbf{q}$ in G_R is connected.*

Thus, flag domains are precisely the G_R-orbits of parabolic subalgebras $\mathbf{q}$ where $(\mathbf{g}_R,\mathbf{q})$ is a compact pair. This suggests the following

Definition. A *canonical flag domain* is the G_R-orbit of a parabolic subalgebra $\mathbf{q}$ where $(\mathbf{g}_R,\mathbf{q})$ is a canonical pair.

Let $D\subset F$ be a flag domain and $\mathbf{q}\in D$. If $\mathbf{q}$ is the Lie algebra of P then the stabiliser of $\mathbf{q}$ is $H=G_R\cap P$. Let the canonical element of $(\mathbf{g}_R,\mathbf{q})$ be ξ, then, as above, the Lie algebra of H is given by $\mathbf{h}=\mathbf{g}_0\cap\mathbf{g}_R$ and setting

$$\mathbf{m}=\sum_{r\neq 0}\mathbf{g}_r\cap\mathbf{g}_R$$

gives an $\mathrm{Ad}(H)$-invariant complement to $\mathbf{h}$ which makes D into a reductive G_R-space.

We define a G_R-invariant complex structure on D by

$$T^{1,0}D=[\mathbf{n}]$$

which is just the restriction to D of the J_1 of F so we denote it by J_1 also.

Finally, we define the *canonical section* $\Xi:D\to\mathbf{g}_R$ by setting $\Xi(\mathbf{q})$ equal to the canonical element of $(\mathbf{g}_R,\mathbf{q})$. This provides a G_R-equivariant diffeomorphism onto the adjoint orbit of ξ in $\mathbf{g}_R$. The pull-back by Ξ of the Kostant-Kirillov-Souriau symplectic form is again the Kähler form of a Kähler metric on D but, in contrast to the compact case, this metric is in general indefinite.

Fibrations over symmetric spaces

We may associate with a flag manifold F a family of inner symmetric spaces, since for any subgroup K with $G^{\tau_\xi}\subset K\subset(G^{\tau_\xi})_0$, G/K is a symmetric space with inner involution τ_ξ.

Again, a flag domain D is associated with a family of non-compact inner symmetric spaces G_R/K but these will only be Riemannian when τ_ξ is a Cartan involution, that is when D is a canonical flag domain. For simplicity, we shall take K to be connected in this chapter.

Notation. Given a flag manifold F or a flag domain D we shall denote the corresponding symmetric space by $N(F)$ or $N(D)$.

We note that by proposition (4.5), any inner symmetric space of compact type with connected stabilizer is $N(F)$ for at least one flag manifold F, while ,by proposition (4.11), any inner symmetric space of non-compact type is $N(D)$ for at least one canonical flag domain D (in this case, stabilisers are automatically connected).

Now let X be a flag manifold or flag domain with $N(X)$ the corresponding symmetric space. Pick $y_0\in X$ with $\Xi(y_0)=\xi$ and let K be as above. Then if $x_0\in N(X)$ has stabilizer K, we have a homogeneous fibration $\pi:X\to N(X)$ with $\pi(y_0)=x_0$. Such homogeneous fibrations are precisely those for which the tangent bundle to the fibres is given by

$$\beta^X(\ker\pi_*) = \sum[\mathfrak{g}_{2r}]$$

and we call them the *canonical fibrations* of X over $N(X)$. Clearly, all our canonical fibrations have the same fibres. We recall from chapter 3 that the points in $N(X)$ with stabiliser K are parametrised by the finite group of isometries $\Sigma_{N(X)}$. Indeed, $\Sigma_{N(X)}$ acts simply transitively on the canonical fibrations by post-composition so that there are precisely $|\Sigma_{N(X)}|$ canonical fibrations of X over $N(X)$. In particular, each canonical flag domain has a unique canonical fibration.

By analogy with the procedures of chapter 2, we may define another almost complex structure on $F(X)$ by reversing the orientation of J_1 on the fibres of the canonical fibrations. This almost complex structure, which we denote by J_2, has $\sqrt{-1}$-eigenbundle

$$\sum_{r<0}[\mathfrak{g}_{2r}]\oplus\sum_{r\geq 0}[\mathfrak{g}_{2r+1}] = [\mathbf{k}^{\mathbf{C}}\cap\bar{\mathbf{n}}]\oplus[\mathbf{p}^{\mathbf{C}}\cap\mathbf{n}] .$$

We now have the following theorem whose proof will be deferred until chapter 5 when a stronger result will be proved.

Theorem 4.14. *Let X be a flag manifold or canonical flag domain. Then any canonical fibration $\pi:X\to N(X)$ is a twistor fibration.*

C. Geometry of flag spaces

Super-horizontal maps

Let $\pi:X\to N(X)$ be a canonical fibration and let $\mathbf{H}=T^{1,0}_{J_1}X\cap T^{1,0}_{J_2}X$ be the bundle of horizontal (1,0) vectors. It is natural to ask whether $\mathbf{H}$ is a holomorphic sub-bundle of $T^{1,0}_{J_1}X$. In general this will not be the case. However, we distinguish a smaller sub-bundle of $\mathbf{H}$ which is holomorphic. For this we note that

$$\mathbf{H} = \sum_{r\geq 0}[\mathfrak{g}_{2r+1}] .$$

Definition. The sub-bundle $[\mathfrak{g}_1]$ of $\mathbf{H}$ is called the *super-horizontal distribution.*

Proposition 4.15. *The super-horizontal distribution is a holomorphic sub-bundle of $T^{1,0}_{J_1}X$.*

Proof. It is well-known that a sub-bundle $V\subset T^{1,0}$ is holomorphic if and only if

$$[C^\infty(V), C^\infty(T^{0,1})]\subset C^\infty(V\oplus T^{0,1}) .$$

In our situation, if Y, Z are vector fields on X, from (1.4) we have

$$\beta^X(D_YZ)-\beta^X(D_ZY)+p[\beta^X(Y),\beta^X(Z)] = \beta^X([X,Y])$$

where $p:\underline{\mathfrak{g}}\to\sum_{r\neq 0}[\mathfrak{g}_r]$ is orthoprojection. Since each $[\mathfrak{g}_i]$ is D-parallel and

$$p[\mathfrak{g}_1, \sum_{r<0}\mathfrak{g}_r]\subset\sum_{r<0}\mathfrak{g}_r ,$$

our proposition now follows. □

Remarks. (i) From theorem (3.4) and formula (3) we conclude that for $r>1$, $[\mathfrak{g}_{-1}, \mathfrak{g}_r]\neq 0$ so that

$\mathbf{H}$ itself is not holomorphic unless $\mathbf{H} = [\mathbf{g}_1]$. Indeed the super-horizontal distribution is the largest invariant holomorphic sub-bundle of $\mathbf{H}$.

(ii) Some other authors (e.g Bryant [11]) require in their definition of a twistor space that $\mathbf{H}$ be holomorphic. Thus our definition is less restrictive.

Definition. A map $\varphi: M \to X$ is *super-horizontal* if $d\varphi(TM^{\mathbf{C}}) \subset [\mathbf{g}_1] \oplus [\mathbf{g}_{-1}]$.

The following theorem shows that among the homogeneous fibrations the canonical fibrations are dual to the holomorphic ones.

Theorem 4.16. *Let X be a flag manifold (resp. flag domain) and $\pi: X \to N$ a homogeneous fibration onto a Riemannian symmetric space of semi-simple type. Then there is a flag manifold (resp. flag domain) Y and homogeneous fibrations $\pi_1: X \to Y$, $\pi_2: Y \to N$ such that*

(i) π_1 is holomorphic with respect to J_1 and maps the super-horizontal distribution of X onto that of Y;

(ii) π_2 is a canonical fibration;

(iii) $\pi = \pi_2 \circ \pi_1$.

Thus any homogeneous fibration factors as a holomorphic fibration followed by a canonical one.

Proof. For simplicity, we shall suppose that $X = G^{\mathbf{C}}/P = G/H$ is a flag manifold, the argument for flag domains being similar. We let $\mathbf{q}$ be the Lie algebra of P with nilradical $\mathbf{n}$. Let $N = G/K$ with $H \subset K$ so that π is just the coset fibration $G/H \to G/K$ and let τ be the involution at eK with symmetric decomposition $\mathbf{g} = \mathbf{k} \oplus \mathbf{p}$.

Since $\mathbf{h} \subset \mathbf{k}$, the canonical element of $\mathbf{q}$ is in $\mathbf{k}$ and hence both $\mathbf{q}$ and $\mathbf{n}$ are τ-stable. We set $\mathbf{n}_{\mathbf{p}} = \mathbf{n} \cap \mathbf{p}^{\mathbf{C}}$ and observe that $\mathbf{n}_{\mathbf{p}}$ is maximal isotropic in $\mathbf{p}^{\mathbf{C}}$ while $[\mathbf{n}_{\mathbf{p}}, \mathbf{n}_{\mathbf{p}}] \subset \mathbf{n}$ is isotropic. Thus, by theorem (4.8), $\mathbf{n}' = \mathbf{n}_{\mathbf{p}} \oplus [\mathbf{n}_{\mathbf{p}}, \mathbf{n}_{\mathbf{p}}]$ is the nilradical of a parabolic subalgebra $\mathbf{q}'$ for which τ is canonical. Clearly $\mathbf{n}' \subset \mathbf{n}$ and taking polars gives $\mathbf{q} \subset \mathbf{q}'$. Let P' be the normaliser of $\mathbf{q}'$ in $G^{\mathbf{C}}$ and $H' = G \cap P'$. We set $Y = G^{\mathbf{C}}/P'$ and then, since $H' \subset H \subset K$, we let $\pi_1: X \to Y$ and $\pi_2: Y \to Z$ be the coset fibrations. Clearly $\pi = \pi_2 \circ \pi_1$ and π_2 is a canonical fibration.

It remains to prove (i). Let β^X, β^Y be the Maurer-Cartan forms of X and Y. From lemma (1.8) we have

$$\pi_1{}^*\beta^Y = p\beta^X ,$$

where $p: \underline{\mathbf{g}}^{\mathbf{C}} \to [\mathbf{n}' \oplus \bar{\mathbf{n}}']$ is orthoprojection. Now,

$$\pi_1{}^*\beta^Y(T^{1,0}X) = p\beta^X(T^{1,0}X) = p[\mathbf{n}] \subset [\mathbf{n}'] ,$$

since $p\mathbf{n} \subset p\mathbf{q}' \subset \mathbf{n}'$. Thus π_1 is holomorphic.

Lastly, we consider the super-horizontal distributions. Let $\mathbf{g}_i$ be the $\sqrt{-1}i$-eigenspace of the canonical element of $\mathbf{q}$. Then the super-horizontal distribution of X is $[\mathbf{g}_1]$ and, using formula (4), it suffices to show that

$$p\mathbf{g}_1 \perp [\bar{\mathbf{n}}', \bar{\mathbf{n}}'] .$$

For this, write $\mathbf{g}_1 = \mathbf{g}_1 \cap \mathbf{k}^{\mathbf{C}} \oplus \mathbf{g}_1 \cap \mathbf{p}^{\mathbf{C}}$. Since $\mathbf{g}_0 \subset \mathbf{k}^{\mathbf{C}}$, we check that

$$[\mathbf{g}_1 \cap \mathbf{k}^{\mathbf{C}}, \bar{\mathbf{n}}_{\mathbf{p}}] \subset \bar{\mathbf{n}}_{\mathbf{p}}$$

whence $\mathfrak{g}_1 \cap \mathfrak{k}^{\mathbb{C}} \subset \mathfrak{q}' \cap \overline{\mathfrak{q}}'$ and so $p\mathfrak{g}_1 = \mathfrak{g}_1 \cap \mathfrak{p}^{\mathbb{C}}$. Finally, since $\mathfrak{g}_1 \perp [\overline{\mathfrak{n}}, \overline{\mathfrak{n}}]$, we see that $\mathfrak{g}_1 \cap \mathfrak{p}^{\mathbb{C}} \perp [\overline{\mathfrak{n}}', \overline{\mathfrak{n}}']$ and the proof is complete. □

Remark. If $\pi: X \to N$ is a canonical fibration then $X = Y$ and $\pi = \pi_2$. Conversely, if N is Hermitian symmetric and π is holomorphic then $N = Y$ and $\pi = \pi_1$.

Corollary 4.17. *Let M be an almost Hermitian cosymplectic manifold and $\pi: X \to N$ a homogeneous fibration of a flag manifold or flag domain onto a Riemannian symmetric space of semi-simple type. If $\psi: M \to X$ is a super-horizontal holomorphic map then $\pi \circ \psi: M \to N$ is a harmonic map.*

Proof. In the notation of (4.16), $\pi_1 \circ \psi: M \to Y$ is J_1-holomorphic and super-horizontal and so, in particular, is J_2-holomorphic. Thus $\pi_2 \circ \pi_1 \circ \psi$ is harmonic since π_2 is a twistor fibration by (4.14). □

Remark. The importance of (4.17) is that, by (4.15), super-horizontality is a purely holomorphic condition. Thus, if M is a Riemann surface, we may find harmonic maps by solving a holomorphic o.d.e. (c.f. [11]). However, even when $M = S^2$ and $N = G_{r,n}$, this method does not produce *all* harmonic maps unless $r = 1$.

Various special cases of (4.17) have been known to a number of authors:
(i) In the case $G = \mathbf{SU}(n)$ and M a Riemann surface, this result is due to Erdem-Wood [35] and, independently, to Zakrzewski [86]. Here the symmetric spaces N are the complex Grassmannians.
(ii) For G compact semi-simple, P a Borel subgroup of $G^{\mathbb{C}}$ and M a Riemann surface, the result is due to Bryant [11].

Super-horizontal maps have also arisen in other contexts. In [72], Uhlenbeck defines an S^1-action on harmonic maps of S^2 into a compact Lie group G. It can be shown that the fixed points of this action are precisely the homogeneous projections of super-horizontal maps into flag manifolds. (Here the symmetric space N is totally geodesically immersed in G via (1.9).)

Lastly, we remark that among the flag domains are the period matrix domains of Griffiths (see section D below). In this context, the super-horizontal distribution is the distribution of vectors that satisfy the infinitesimal bilinear relation (see below and [38]).

Height

The geometry of a flag manifold clearly depends on the eigenvalues of its canonical element. This motivates the following

Definition. A flag manifold or flag domain with canonical section Ξ has *height* r if r is the largest eigenvalue of $\sqrt{-1}\,\mathrm{ad}\Xi$ (or equivalently, if the nilradicals of the corresponding parabolic subalgebras are r-step nilpotent).

Examples. (i) If X has height one, then X and $N(X)$ coincide and $N(X)$ is a Hermitian symmetric space: these are the only spaces which are simultaneously flag manifolds and symmetric spaces.

(ii) The height two flag manifolds admit several characterisations: from (4.15) and the remark following it, we see these are precisely the non-symmetric flag manifolds for which the horizontal (1,0) vectors form a holomorphic distribution (for J_1). Alternatively, Salamon has characterised them as flag manifolds which admit a homogeneous metric which is (1,2)-symplectic for J_2 (i.e if ω is the corresponding Kähler form, then $d\omega^{(1,2)}=0$). This, in turn, follows from general results of Wolf-Gray [81] since such flag manifolds are 3-symmetric spaces. (For details, see [62]).

Indeed, if X has height r, then setting $\tau_x = \mathrm{Adexp}2\pi\Xi(x)/r$ gives X the structure of a Riemannian (regular) r-symmetric space (c.f. [50]).

In another direction, we remark that the fibres of the canonical fibrations are themselves flag manifolds holomorphically embedded in X. Moreover, if $r \leq 4$, the fibres are Hermitian symmetric spaces totally geodesically embedded in X. For a detailed study of such fibrations of 4-symmetric spaces the Reader is referred to [47].

Classification questions

It would be useful to know which flag manifolds fibred canonically over which symmetric spaces. Bryant [11] and Salamon [62] have settled this question for flag manifolds of height one or two by checking cases. Further, given F, combinatorial arguments on the Dynkin diagram of G enable one to identify $N(F)$ by examining the orbit of the canonical element modulo the root lattice under the Weyl group. In this way, for example, one may show that of the 63 non-trivial flag manifolds of $\mathbf{E}_6$, 27 fibre over $\mathbf{E}_6/\mathbf{U}(1)\times\mathbf{Spin}(10)$ and 36 over $\mathbf{E}_6/\mathbf{Sp}(1)\mathbf{SU}(6)$.

However, we know of no general method for settling this question without case by case checking. We give some partial results below under the simplifying assumption that G is simple.

Firstly we have

Proposition 4.18. *Let G be simple and N an inner symmetric G-space. Then there is a flag manifold of height no greater than two fibring canonically over N.*

Proof. Recall from (3.17) that the inner involution on N is conjugate to one of the form $\mathrm{Adexp}\pi\xi_i$ where α_i has coefficient one or two in the highest root θ. Such an involution is canonical for the parabolic subalgebra $\mathbf{q}_{\{i\}}$ which has canonical element ξ_i. The corresponding flag manifold then fibres canonically over N and is easily seen to have height $\theta(\xi_i)$ which is just the coefficient of α_i in θ. □

In case that N is either S^{2n} or $\mathbf{C}P^n$, we shall show below that *all* the corresponding flag manifolds have height no greater than two.

In another direction we have

Proposition 4.19. *Let N be a simply connected compact irreducible inner symmetric space. Then $\pi_2(N)=0$ if and only if only flag manifolds of even height fibre canonically over N.*

Proof. Let $\pi : F \to N$ be a canonical fibration, $\mathbf{q}$ the parabolic subalgebra at some $x \in F$ and $\mathbf{k}\oplus\mathbf{p}$ the symmetric decomposition at $\pi(x)$. We fix a torus in $\mathbf{q}\cap\mathbf{g}$ (necessarily fundamental for the involution at $\pi(x)$) so that $\mathbf{q}=\mathbf{q}_I$ for some set of simple roots. Let θ be the corresponding highest root. F has height r if and only if $\theta \in \mathbf{g}_r$ so that if r is odd, $\theta \in I_{\mathbf{p}}$ and by (3.19) $\pi_2(N) \neq 0$ since θ is long.

Conversely, suppose that $\pi_2(N) \neq 0$ and fix a fundamental torus at some point. Then again by (3.17) we have a long root in $I_{\mathbf{p}}$ which after conjugation by the Weyl group we may assume to be the highest root for some set of simple roots $\{\alpha_1,\ldots,\alpha_l\}$. We set

$$I = \{i : \mathbf{g}^{\alpha_i} \subset \mathbf{p}\}$$

and argue as in (4.5) to conclude that $\mathbf{q}_I$ is canonical. Let F be the corresponding flag manifold which fibres canonically over N. If the height of F is r then since $\theta \in \mathbf{g}_r$ we conclude that r is odd. This completes the proof. □

Finally, we remark that it sometimes occurs that two parabolic subgroups P_1, P_2 may have the same real part so that the corresponding flag manifolds F_1 and F_2 coincide as G-spaces and further that the J_2 structures on the F_i coincide even though $N(F_1) \neq N(F_2)$. In this circumstance, $F_i = F$ fibres canonically over two different symmetric spaces to give two twistor fibrations of (F, J_2).

Example. The Dynkin diagram of $\mathbf{E}_6$ is

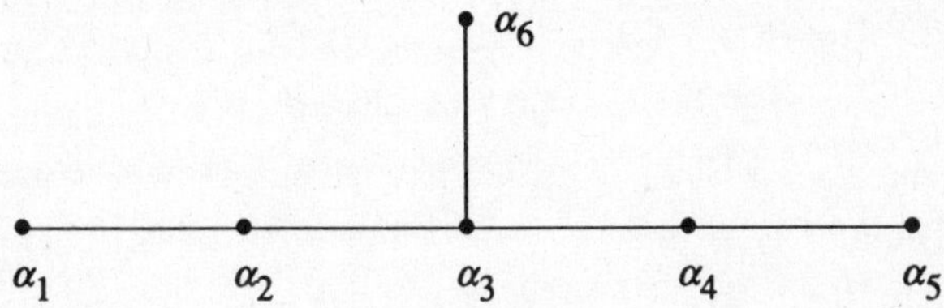

and taking $I_1 = \{1,2,4,5\}$ and $I_2 = \{1,2,5,6\}$ gives flag manifolds with common real description $\mathbf{E}_6/\mathbf{U}(1)\times\mathbf{U}(1)\times\mathbf{U}(1)\times\mathbf{U}(3)$ (and common J_2!) of which there are canonical twistor fibrations over both inner symmetric $\mathbf{E}_6$-spaces.

The Reader is referred to [14] for details of when this circumstance occurs (see also [15] for the case $G = \mathbf{SU}(n)$).

D. Topology of flag manifolds

In this section, we apply the methods of chapter 3 to discuss the topology of flag manifolds. In particular, we study the cohomology class represented by the J_1 Kähler form that we have introduced. It is well-known that flag manifolds are Hodge manifolds so that this Kähler form represents an *integral* cohomology class but in fact we can say much more. We shall show that it represents the Chern class of a natural vector bundle on the flag manifold and derive an explicit formula for that Chern class in terms of the generators of $H^2(F,\mathbf{Z})$.

In chapter 8, these results will be applied to the study of the energy of harmonic 2-spheres in Lie groups.

So let F be a G-flag manifold with canonical section Ξ and Maurer-Cartan form β. Of course, we have the decomposition of $\underline{\mathbf{g}}^{\mathbf{C}}$ into eigen-bundles of $\mathrm{ad}\Xi$:

$$\underline{\mathbf{g}}^{\mathbf{C}} = \sum[\mathbf{g}_r], \qquad [\mathbf{q}] = \sum_{r\geq 0}[\mathbf{g}_r], \qquad [\mathbf{n}] = \sum_{r>0}[\mathbf{g}_r] .$$

The Kähler form ω is given by

$$\omega = -\tfrac{1}{2}(\Xi,[\beta\wedge\beta]) ,$$

where the inner product is the negative of the Killing form of $\mathbf{g}$.

π_2 of flag manifolds

It is well-known that flag manifolds are simply connected [6] so that without loss of generality we may take G to be simply connected also. Thus we may apply the methods of chapter 3 to determine π_2 of flag manifolds.

Recall that if $\mathbf{g}$ is semi-simple and $\mathbf{t}\subset\mathbf{g}$ is a maximal torus, we define the lattice $\Lambda_{\mathbf{g}}(\mathbf{t})$ as follows: fix an invariant inner product on $\mathbf{g}$ and, for $\lambda\in(\mathbf{t}^{\mathbf{C}})^*$, define $H_\lambda\in\mathbf{t}^{\mathbf{C}}$ by

$$(H_\lambda,\xi) = \lambda(\xi)$$

for $\xi\in\mathbf{t}^{\mathbf{C}}$. For $\alpha\in\Delta=\Delta(\mathbf{g}^{\mathbf{C}},\mathbf{t}^{\mathbf{C}})$, set

$$\eta_\alpha = 2\sqrt{-1}H_\alpha/(\alpha,\alpha)\in\mathbf{t}$$

and put $\Lambda_{\mathbf{g}}(\mathbf{t})=\mathrm{span}_{\mathbf{Z}}\{\eta_\alpha:\alpha\in\Delta\}$. We recall that if $\Phi\subset\Delta$ is the set of simple roots for some Weyl chamber then $\{\eta_\alpha:\alpha\in\Phi\}$ is a basis for $\Lambda_{\mathbf{g}}(\mathbf{t})$.

Suppose now that G/H is a homogeneous space with G semi-simple and simply-connected and K connected. If $\mathbf{t}\subset\mathbf{g}$ is a maximal torus for which $\mathbf{t}\cap[\mathbf{h},\mathbf{h}]$ is maximal abelian in $[\mathbf{h},\mathbf{h}]$, we apply (3.13), (3.14) and (3.15) to conclude that

$$\pi_2(G/H) = \Lambda_{\mathbf{g}}(\mathbf{t})/\Lambda_{[\mathbf{h},\mathbf{h}]}(\mathbf{t}\cap[\mathbf{h},\mathbf{h}])\ .$$

Let us apply this to the case in hand. We fix a base-point $x\in F$ with stabiliser H and $[\mathbf{q}]_x=\mathbf{q}$. Now choose a maximal torus $\mathbf{t}\subset\mathbf{g}\cap\mathbf{q}$ and a set of simple roots $\alpha_1,\ldots,\alpha_l$ so that

$$\mathbf{q} = \mathbf{q}_I$$

for some $I\subset\{1,\ldots,l\}$. It is straightforward to check that $\mathbf{t}\cap[\mathbf{h},\mathbf{h}]$ is maximal abelian in $[\mathbf{h},\mathbf{h}]$ and that $\{\alpha_j:\ j\notin I\}$ is a set of simple roots for $[\mathbf{h},\mathbf{h}]^{\mathbf{C}}$. Thus we conclude

Proposition 4.20 $\pi_2(F)\cong$ *$\mathrm{span}_{\mathbf{Z}}\{\eta_{\alpha_i}:\ i\in I\}\cong\mathbf{Z}^{|I|}$.*

Further, we can produce a set of homogeneous generators for $\pi_2(F)$ by following the methods of chapter 3, section E. Indeed, let $\mathbf{g}^\alpha\in\mathbf{n}$, then $\mathbf{g}^\alpha\oplus\mathbf{g}^{-\alpha}\oplus[\mathbf{g}^\alpha,\mathbf{g}^{-\alpha}]$ is a Lie algebra isomorphic to $\mathbf{sl}(2,\mathbf{C})$ and this is the complexification of a subalgebra $\mathbf{s}_\alpha$ of $\mathbf{g}$ which is isomorphic to $\mathbf{su}(2)$. We let S_α be the corresponding copy of $\mathbf{SU}(2)$ and $U_\alpha\subset S_\alpha$ the connected subgroup with Lie algebra $\mathbf{u}_\alpha=[\mathbf{g}^\alpha,\mathbf{g}^{-\alpha}]\cap\mathbf{g}\subset\mathbf{h}$. Clearly, U_α is a copy of $\mathbf{U}(1)$ so that the homogeneous space S_α/U_α is a 2-sphere. Moreover, S_α/U_α is a flag manifold with $[\mathbf{g}^\alpha]$ as nil-radical bundle. We denote the corresponding Kähler form on S^2 by ω_α.

Now let $i_\alpha:S_\alpha\to G$ be the homomorphism to which the inclusion $\mathbf{s}_\alpha\subset\mathbf{g}$ exponentiates. Then, as before, since $i_\alpha(U_\alpha)\subset H$, we can define an i_α-equivariant immersion $\varphi_\alpha:S^2\to F$.

Proposition 4.21. *(i) φ_α is totally geodesic with respect to the canonical connections on S^2 and F.*

(ii) φ_α represents the class $[\eta_\alpha]\in\pi_2(F)$.

(iii) φ_α is J_1-holomorphic.

(iv) With $n_I(\alpha)$ defined as in (4.1),

$$\varphi_\alpha{}^*\omega = \frac{n_I(\alpha)}{2|\alpha|^2}\omega_\alpha\ .$$

Proof. Assertions (*i*) and ($\sqrt{-1}$) are proved exactly as in (3.28). For (*iii*), let β^α denote the Maurer-Cartan form of S_α/U_α. Then, by (1.8), we have

$$\varphi_\alpha{}^*\beta = i_\alpha \circ \beta^\alpha \tag{6}$$

from which it is clear that φ_α is holomorphic.

Now we turn to the relationship between ω_α and ω. First let us consider the Killing forms B_α and B on $\mathbf{s}_\alpha$ and $\mathbf{g}$ respectively. Since $\mathbf{s}_\alpha$ is simple, B_α is proportional to B on $\mathbf{s}_\alpha$ by Schur's Lemma. Further, a simple calculation with $H_\alpha \in [\mathbf{g}^\alpha, \mathbf{g}^{-\alpha}]$ defined as above (with respect to B), we have

$$B_\alpha(H_\alpha, H_\alpha) = 2|\alpha|^2 B(H_\alpha, H_\alpha)$$

so that on $\mathbf{s}_\alpha$

$$B_\alpha = 2|\alpha|^2 B\ .$$

Now let us compare canonical elements ξ_α, ξ at the identity coset: since $\mathbf{g}^\alpha \subset \mathbf{g}_{n_I(\alpha)}$ we see that on $\mathbf{s}_\alpha$,

$$\mathrm{ad}\xi = n_I(\alpha)\mathrm{ad}\xi_\alpha\ .$$

Thus, at the identity coset,

$$\begin{aligned}\varphi_\alpha{}^*\omega &= \tfrac{1}{2}B(\Xi, [\varphi_\alpha{}^*\beta \wedge \varphi_\alpha{}^*\beta]) \\ &= \tfrac{1}{2}n_I(\alpha)B(\Xi_\alpha, [\beta^\alpha \wedge \beta^\alpha]) \\ &= \tfrac{1}{2}\frac{n_I(\alpha)}{2|\alpha|^2}B_\alpha(\Xi_\alpha, [\beta^\alpha \wedge \beta^\alpha]) = \frac{n_I(\alpha)}{2|\alpha|^2}\omega_\alpha\ .\end{aligned}$$

This last formula now holds everywhere by equivariance and we are done. □

Corollary 4.22. $\pi_2(F)$ *is generated by superhorizontal holomorphic curves.*

Proof. In fact, for $i \in I$, φ_{α_i} is clearly superhorizontal by (6). □

Kähler forms and cohomology

We now show that ω represents the Chern class of a certain natural vector bundle on F.

Proposition 4.23. $[\omega] = -4\pi \sum_{r>0} r c_1([\mathbf{g}_r])\ .$

Proof. Let D be the canonical connection of F. Then each $[\mathbf{g}_r]$ is invariant and hence D-parallel. The curvature of D is given by (1.4) as

$$R = -\tfrac{1}{2}\mathrm{ad}\pi_0[\beta \wedge \beta]$$

where $\pi_0 : \underline{\mathbf{g}}^{\mathbf{C}} \to [\mathbf{g}_0]$ is orthoprojection. Thus the Chern-Weil theory (c.f. Wells [77]) tells us that the Chern class of $[\mathbf{g}_r]$ is given by

$$c_1([\mathbf{g}_r]) = \frac{\sqrt{-1}}{2\pi}\mathrm{trace}_{[\mathbf{g}_r]}R = -\frac{\sqrt{-1}}{4\pi}\mathrm{trace}_{[\mathbf{g}_r]}\mathrm{ad}\pi_0[\beta \wedge \beta]\ .$$

However,

$$\omega = -\tfrac{1}{2}(\Xi, [\beta \wedge \beta]) = -\tfrac{1}{2}(\Xi, \pi_0[\beta \wedge \beta])\ ,$$

which, by definition of the Killing form, can be written as

$$\tfrac{1}{2}\mathrm{trace}_{\mathbf{g}^{\mathbf{C}}}\,\mathrm{ad}(\pi_0[\beta\wedge\beta])\circ\mathrm{ad}\Xi$$

and since $\mathrm{ad}\Xi = \sqrt{-1}r$ on $[\mathbf{g}_r]$, this is just

$$\frac{\sqrt{-1}}{2}\sum_r r\,\mathrm{trace}_{[\mathbf{g}_r]}\mathrm{ad}\pi_0[\beta\wedge\beta] = -2\pi\sum_r r c_1([\mathbf{g}_r]) \ .$$

Finally, $\overline{\mathbf{g}_r} = \mathbf{g}_{-r}$ so that $c_1([\mathbf{g}_{-r}]) = -c_1([\mathbf{g}_r])$ and the conclusion follows. □

Recall that the i-th term in the central descending series of $\mathbf{n}$ is given by

$$\mathbf{n}^{(i)} = \sum_{r\geq i}\mathbf{g}_r \ ,$$

so that (4.23) may be rephrased as

Corollary 4.24. $[\omega] = -4\pi c_1(\bigoplus_i[\mathbf{n}^{(i)}])$.

In particular, returning to the root spheres S_α/U_α, we see that

$$\int_{S^2}\omega_\alpha = -4\pi c_1(T^{1,0}S^2)[M] = -8\pi \ ,$$

so that, by (4.21iv),

$$\int_{S^2}\varphi_\alpha\omega = -\frac{4\pi}{|\alpha|^2}n_I(\alpha) \ .$$

Since F is simply connected, $\pi_2(F)\cong H_2(F,\mathbf{Z})$ so that, by (4.20) and (4.21ii), we may take $\{\varphi_{\alpha_i}: i\in I\}$ as a basis for $H_2(F,\mathbf{Z})$. Let $\{\omega_i: i\in I\}$ be the dual basis in cohomology. We can now conclude

Proposition 4.25. $[\omega] = -4\pi\sum_{i\in I}\dfrac{\omega_i}{|\alpha_i|^2}$.

Example. Let F be an irreducible Hermitian symmetric space with θ the highest root of $\mathbf{g}^{\mathbf{C}}$. Then F is a height one flag manifold and $I=\{i_0\}$ with α_{i_0} long. Thus ω_{i_0} generates $H^2(F,\mathbf{Z})$ and

$$\omega = -\frac{4\pi}{|\theta|^2}\omega_{i_0} \ .$$

E. Examples

Let us illustrate this general theory of flag spaces by considering two examples.

$\mathrm{SL}(n,\mathbf{C})$ *flag manifolds*

Let $I=\{i_1<\ldots<i_r=n\}\subset\{1,\ldots,n\}$ be a multi-index. A *flag of index* I is a filtration of $\mathbf{C}^n$ by subspaces V_i

$$V_1\subset V_2\subset\ldots\subset V_r = \mathbf{C}^n$$

with $\dim V_j = i_j$.

The parabolic subalgebras of $\mathbf{sl}(n,\mathbf{C})$ are precisely those of the form

$$\{T\in\mathbf{sl}(n,\mathbf{C}):\ TV_j\subset V_j \quad \forall j\}$$

for some flag. The corresponding parabolic subgroups are

$$\{T\in\mathbf{SL}(n,\mathbf{C}):\ TV_j\subset V_j \quad \forall j\}\ .$$

Remark. If $I=\{1<\ldots<n\}$ (respectively $I=\{n\}$), then the parabolic subalgebra is a Borel subalgebra (respectively $\mathbf{sl}(n,\mathbf{C})$).

Fixing a parabolic subalgebra $\mathbf{q}$ with flag $f=V_1\subset\ldots\subset V_r$, let $\mathbf{n}^{(i)}$ denote the ith step of the central descending series of the nilradical of $\mathbf{q}$. It is easy to check that

$$\mathbf{n}^{(i)} = \{T\in\mathbf{sl}(n,\mathbf{C}):\ TV_j\subset V_{j-i} \quad \forall j\}$$

where we set $V_j=\{0\}$ for $j\leq 0$.

Finding a compact real form of $\mathbf{sl}(n,\mathbf{C})$ amounts to fixing a Hermitian metric on $\mathbf{C}^n$. The real form is then $\mathbf{su}(n)$ and we may define mutually orthogonal subspaces $E_1,\ldots,E_r$ by

$$E_i = V_i\cap V_{i-1}^{\perp}\ .$$

Then

$$\mathbf{q}\cap\mathbf{su}(n) = \{T\in\mathbf{su}(n):\ TE_j\subset E_j \quad \forall j\} \cong \mathbf{s}(\mathbf{u}(i_1)\times\ldots\times\mathbf{u}(n-i_{r-1}))\ .$$

Now define $\xi\in\mathbf{q}\cap\mathbf{su}(n)$ by

$$\xi v = \sqrt{-1}(N-i)v$$

for $v\in E_i$, where $N=\sum\dim E_i/n$ is chosen to force ξ to have trace zero. Then the $\sqrt{-1}i$-th eigenspace of $\mathrm{ad}\xi$ is easily shown to be

$$\mathbf{g}_i = \{T\in\mathbf{sl}(n,\mathbf{C}):\ TE_j\subset E_{j-i} \quad \forall j\}$$

and so we conclude that ξ is the canonical element of $\mathbf{q}$.

As for the canonical involution τ_ξ, set $E_+=\sum E_{2j}$, $E_-=\sum E_{2j+1}$ and then we see that τ_ξ has symmetric decomposition

$$\mathbf{k} = \{T\in\mathbf{su}(n):\ TE_+\subset E_+,\ TE_-\subset E_-\} \qquad \mathbf{p} = \{T\in\mathbf{su}(n):\ TE_+\subset E_-,\ TE_-\subset E_+\}\ .$$

Globalising now, observe that $\mathbf{SL}(n,\mathbf{C})$ acts transitively on all flags of fixed index I so that our flag manifolds are parametrised by the index I with F_I being realised as the set of all flags of index I. We now see that the symmetric spaces $N(F_I)$ are just the complex Grassmannians $G_{r,n}$ (c.f. Chapter 1) with $r=\dim E_+$. In this setting, the canonical fibration is given by

$$f\mapsto E_+ = \sum E_{2i}\ .$$

We note that the height of F_I is $r-1$.

A flag manifold F admits various tautological sub-bundles $\mathbf{V}_i$, $\mathbf{E}_i$ of $\underline{\mathbf{C}}^n$ whose fibres at $V_1\subset\ldots\subset V_r$ are V_i and $V_{i-1}^{\perp}\cap V_i$ respectively. These bundles are homogeneous for $\mathbf{SL}(n,\mathbf{C})$ and $\mathbf{SU}(n)$ respectively. Further, since

$$\mathbf{g}_i.E_j\perp E_j$$

for all non-zero i and all j, we conclude from (1.5) that the Maurer-Cartan form of F is just the sum of the second fundamental forms of the $\mathbf{E}_i$.

Now let $\varphi: M \to F$ be a map, then we have a smooth orthogonal decomposition of the trivial $\mathbf{C}^n$ bundle over M:

$$M \times \mathbf{C}^n = \varphi^{-1}\mathbf{E}_1 \oplus \ldots \oplus \varphi^{-1}\mathbf{E}_r \ .$$

Conversely, such a decomposition defines a map into a flag manifold. If M is complex, we see that φ is holomorphic with respect to J_1 if and only if

$$\partial_X C^\infty(\varphi^{-1}\mathbf{E}_i) \subset C^\infty(\sum_{j \geq i} \varphi^{-1}\mathbf{E}_j)$$

for $X \in C^\infty(T^{0,1}M)$, or, equivalently, if each $\varphi^{-1}\mathbf{V}_i^\perp$ is a holomorphic sub-bundle of $M \times \mathbf{C}^n$. Further, φ is super-horizontal holomorphic precisely when, in addition,

$$\partial_X C^\infty(\varphi^{-1}\mathbf{V}_i^\perp) \subset C^\infty(\varphi^{-1}\mathbf{V}_{i-1}^\perp)$$

for $X \in C^\infty(T^{1,0}M)$, (c.f. [35]).

Lastly, if $N(F_I) = \mathbf{C}P^{n-1}$ then $\dim E_+$ is 1 or $n-1$ so that $I = \{i, i+1, n\}$, $\{1, n\}$ or $\{n-1, n\}$ from which we conclude that

Proposition 4.26. *If F is an* $\mathbf{SU}(n)$*-flag manifold and $N(F)$ is $\mathbf{C}P^{n-1}$ then F has height not exceeding two.*

In fact, the height one flag manifolds in this case are just copies of $\mathbf{C}P^{n-1}$ with holomorphic or anti-holomorphic canonical fibration while the height two flag manifolds are the $H_{r,s}$ spaces of Eells-Wood [34].

Remark. For more details on $\mathbf{SU}(n)$ flag manifolds and their twistor fibrations the Reader is referred to [15] with the warning that the J_i almost complex structures defined therein differ from ours by sign.

Period matrix domains

Our second example comes from Griffiths' study of variation of Hodge structure [38]. In that work the classifying spaces for Hodge structures or *period matrix domains* are a family of flag domains which we now describe.

We begin with the following data: let $n \in \mathbf{N}$, V a complex vector space and Q a non-degenerate bilinear form on V satisfying

$$Q(v,w) = (-1)^n Q(w,v) \ .$$

For W a subspace of V, we let $W^\perp$ denote the polar of W with respect to Q.

An *isotropic flag* is a filtration of V by subspaces

$$0 \neq V_0 \subseteq \ldots \subseteq V_{n-1} \subset V_n = V$$

with the property that

$$V_j^\perp = V_{n-j-1} \ . \tag{7}$$

If $\dim V_j = h_j$, we say that the flag has index $h_0 \leq \ldots \leq h_{n-1} < h_n$. Note that in contrast to the previous example we do not exclude the possibility that $V_j = V_{j+1}$ for some $j < n-1$.

Remark. The condition (7) is the *first Riemann bilinear relation.*

Let $G^{\mathbf{C}}$ be the complex orthogonal group of Q: thus $G^{\mathbf{C}} = \mathbf{SO}(h_n,\mathbf{C})$ or $\mathbf{Sp}(h_n,\mathbf{C})$ depending on the parity of n. $G^{\mathbf{C}}$ acts transitively on the set of all isotropic flags of a given index and the stabiliser of each such flag is a parabolic subgroup of $G^{\mathbf{C}}$. Thus the set of all isotropic flags of a fixed index becomes a flag manifold.

Let us fix an index $h_0 \leq \ldots \leq h_{n-1} < h_n$ and denote the corresponding flag manifold by F. Let $f \in F$, $F = V_0 \subset \ldots \subset V_n$, have stabiliser P. To understand the structure of $\mathbf{q}$, the Lie algebra of P, we define a reduced index $i_0 < \ldots < i_r$ consisting of the distinct elements of $\{h_j\}$ and a reduced flag

$$0 \neq W_0 \subset \ldots \subset W_r = V$$

with $\dim W_k = i_k$, by setting $W_k = V_j$ for some j with $i_k = h_j$. Thus, in the reduced flag each distinct subspace in f is counted only once. Now the central descending series of the nilradical of $\mathbf{q}$ is given by

$$\mathbf{n}^{(i)} = \{T \in \mathbf{g}^{\mathbf{C}} : TW_k \subset W_{k-i} \quad \forall k\}$$

where, as before, we set $W_k = \{0\}$ for $k<0$.

Let us now suppose that there is a real form $V_{\mathbf{R}}$ of V on which Q is real. We denote conjugation with respect to $V_{\mathbf{R}}$ by $v \mapsto \bar{v}$ and define an indefinite Hermitian inner product $(.,.)$ on V by

$$(v,w) = (-\sqrt{-1})^n Q(v,\bar{w}) .$$

We set

$$E_j = \{v \in V_j : (v, V_{j-1}) = 0\} , \qquad F_k = \{v \in W_k : (v, W_{k-1}) = 0\}$$

and let $D \subset F$ be the set of all flags in F which satisfy the *second Riemann bilinear relation*:

(i) $(.,.)$ is non-singular on each V_j;
(ii) $(-1)^j(.,.)$ is positive definite on E_j for all j.

D is an open subset of F and is called a *period matrix domain.*

Let $G_R \subset G^{\mathbf{C}}$ be the subgroup of $G^{\mathbf{C}}$ that preserves $V_{\mathbf{R}}$ and hence $(.,.)$. Then G_R is a simple non-compact Lie group which acts transitively on D. Clearly the stabiliser of a flag in D is compact so that $\mathbf{g}_R$ has inner Cartan involution and D is a flag domain.

We now suppose that $f \in D$ and then we have

$$V_j = \bigoplus_{i \leq j} E_i , \qquad W_k = \bigoplus_{h \leq k} F_h .$$

The $\sqrt{-1}i$-th eigenspace of the canonical element at f is given by

$$\mathbf{g}_i = \{T \in \mathbf{g}^{\mathbf{C}} : TF_k \subset F_{k-i} \quad \forall k\}$$

and, in particular, the anti-holomorphic super-horizontal distribution at f is

$$\{T \in \mathbf{g}^{\mathbf{C}} : TF_k \subset F_{k+1} \quad \forall k\} .$$

In Griffiths' theory of period mappings, an important role is played by the distribution of vectors satisfying the *infinitesimal bilinear relation.* In our setting, this distribution is given at f by

$$\{T \in \mathbf{g}^{\mathbf{C}} : TE_j \subset E_{j+1} \quad \forall j\} .$$

Thus we see that this distribution is a sub-bundle of the anti-holomorphic super-horizontal distribution and coincides with it when $E_i = F_i$ for all i, i.e. when all the V_i are distinct.

Lastly, we consider the canonical involution. We set $F_+ = \sum F_{2k}$, $F_- = \sum F_{2k+1}$ and then it is easily shown that

$$\mathbf{k} = \{T\in\mathbf{g}: TF_+\subset F_+,\ TF_-\subset F_-\}\ , \qquad \mathbf{p} = \{T\in\mathbf{g}: TF_+\subset F_-,\ TF_-\subset F_+\}$$

gives the symmetric decomposition for the canonical involution. This involution will be a Cartan involution if and only if (.,.) is definite on F_+ and on F_-. A necessary condition for this is again that all the V_i are distinct.

In summary, the period matrix domains are flag domains which are canonical if all the components V_i of the flags are distinct. In this case the anti-holomorphic super-horizontal distribution is precisely the bundle of vectors that satisfy the infinitesimal bilinear relation and so the local lifts of period mappings are precisely the super-horizontal anti-holomorphic maps into D.

Appendix: Non-inner involutions and parabolic subalgebras

The theory developed in this chapter provides a natural interaction between parabolic subalgebras and inner involutions. However, for our study of the stability of harmonic 2-spheres in chapter 7, we shall need some facts about the relationship between non-inner involutions and parabolic subalgebras which are collected in this appendix.

For simplicity, we shall assume that $\mathbf{g}$ is a compact semi-simple Lie algebra and τ an involution on $\mathbf{g}$ with symmetric decomposition $\mathbf{g}=\mathbf{k}\oplus\mathbf{p}$.

τ-stable parabolic subalgebras

Let $\mathbf{q}$ be a τ-stable parabolic subalgebra of $\mathbf{g}^{\mathbf{C}}$ with nilradical $\mathbf{n}$ and $\mathbf{h}=\mathbf{q}\cap\mathbf{g}$. Thus

$$\mathbf{q} = \mathbf{h}^{\mathbf{C}}\oplus\mathbf{n}\ .$$

We set $\mathbf{h}_{\mathbf{k}}=\mathbf{h}\cap\mathbf{k}$, $\mathbf{h}_{\mathbf{p}}=\mathbf{h}\cap\mathbf{p}$, $\mathbf{n}_{\mathbf{k}}=\mathbf{n}\cap\mathbf{k}^{\mathbf{C}}$, $\mathbf{n}_{\mathbf{p}}=\mathbf{n}\cap\mathbf{p}^{\mathbf{C}}$ and similarly for $\mathbf{q}_{\mathbf{k}}$ and $\mathbf{q}_{\mathbf{p}}$.

Lemma 4.27. $\mathbf{h}_{\mathbf{k}}$ *contains a maximal torus of* $\mathbf{k}$.

Proof. Since $\mathbf{k}$ is compact, it is reductive and so splits as the direct sum of its centre and (semi-simple) derived algebra: $\mathbf{k}=\mathbf{z}(\mathbf{k})\oplus\mathbf{k}_{ss}$. This decomposition is orthogonal for the Killing form B of $\mathbf{g}$.

If $z\in\mathbf{z}(\mathbf{k})$ and $\xi\in\mathbf{n}_{\mathbf{k}}$, we see that since z and ξ commute, $\mathrm{ad}_{\mathbf{g}}z\circ\mathrm{ad}_{\mathbf{g}}\xi$ is nilpotent and so has zero trace whence $B(\mathbf{z}(\mathbf{k}),\mathbf{n}_{\mathbf{k}})=0$. Thus $\mathbf{z}(\mathbf{k})\subset\mathbf{h}$.

Further, $\mathbf{n}_{\mathbf{k}}\subset\mathbf{k}_{ss}^{\mathbf{C}}$ is nilpotent so that $\mathbf{q}\cap\mathbf{k}_{ss}^{\mathbf{C}}=\mathbf{n}^{\perp}\cap\mathbf{k}_{ss}^{\mathbf{C}}$ is a parabolic subalgebra of $\mathbf{k}_{ss}^{\mathbf{C}}$ by (4.2). Thus $\mathbf{h}\cap\mathbf{k}_{ss}$ contains a maximal torus of $\mathbf{k}_{ss}$ and the lemma follows. □

τ-maximal parabolic subalgebras

Definition. A τ-stable parabolic subalgebra $\mathbf{q}$ is said to be τ-maximal if and only if

(i) $\mathbf{h}_{\mathbf{p}}\subset\mathbf{z}(\mathbf{h})$;

(ii) $\mathbf{n}=\mathbf{n}_{\mathbf{p}}+[\mathbf{n}_{\mathbf{p}},\mathbf{q}_{\mathbf{p}}]$.

Remark. If τ is inner, $\mathbf{h}_{\mathbf{k}}$ contains a maximal torus of $\mathbf{g}$ and so $\mathbf{z}(\mathbf{h})\subset\mathbf{k}$. Thus, in this case, we see that $\mathbf{q}$ is τ-maximal precisely when τ is canonical for $\mathbf{q}$.

It is known that any τ-stable parabolic subalgebra of $\mathbf{g}^{\mathbf{C}}$ contains a τ-stable maximal torus of $\mathbf{g}$. For τ-maximal parabolic subalgebras, we can say more:

Lemma 4.28. *Let* $\mathbf{q}$ *be* τ*-maximal and* $\mathbf{t}\subset\mathbf{h}$ *a* τ*-stable maximal torus. Then* $\mathbf{t}$ *is a fundamental torus with respect to* τ.

Proof. We must show that $\mathbf{t_k}$ is maximal abelian in $\mathbf{k}$ for which it suffices to show that $\mathbf{t_k}$ is maximal abelian in $\mathbf{h_k}$ by (4.27). But if $\mathbf{t_k}'$ were a torus in $\mathbf{h_k}$ strictly containing $\mathbf{t_k}$, then since $\mathbf{t_p}\subset\mathbf{z}(\mathbf{h})$, $\mathbf{t_k}'\oplus\mathbf{t_p}$ would be a torus in $\mathbf{g}$ strictly containing $\mathbf{t}$ which is a contradiction. □

As an application, we see that, since $\mathbf{t_p}=\mathbf{z_p}(\mathbf{t_k})$ by (3.10), we have

$$\mathbf{t_p} = \mathbf{z_p}(\mathbf{h}) = \mathbf{h_p}$$

for τ-maximal $\mathbf{q}$. From this we deduce that $\mathbf{h}^{\mathbf{C}}$ consists solely of $\mathbf{t}^{\mathbf{C}}$ and some $I_{\mathbf{k}}$ root spaces.

We need to guarantee a good supply of τ-maximal parabolic subalgebras. In fact these can always be constructed from a τ-stable positive root system:

Theorem 4.29. *Let* $\mathbf{a_k}\subset\mathbf{k}^{\mathbf{C}}$ *be a CSA for* $\mathbf{k}^{\mathbf{C}}$ *and let* $\mathbf{a_p}$ *be the centraliser of* $\mathbf{a_k}$ *in* $\mathbf{p}^{\mathbf{C}}$. *Then*
(i) $\mathbf{a}=\mathbf{a_k}\oplus\mathbf{a_p}$ *is a CSA for* $\mathbf{g}^{\mathbf{C}}$;
(ii) if $\mathbf{b}$ *is the Borel subalgebra corresponding to a* τ*-stable set of positive roots* $\Delta^+\subset\Delta(\mathbf{g}^{\mathbf{C}},\mathbf{a})$ *with nilradical* $\mathbf{n}$, *then*

$$\mathbf{n}_0 = \mathbf{n_p}+[\mathbf{n_p},\mathbf{b_p}]$$

is the nilradical of a τ*-maximal parabolic subalgebra.*

Proof. Since $\mathbf{a_k}$ is $\mathbf{k}^{\mathbf{C}}$-conjugate to the complexification of a maximal torus of $\mathbf{k}$, (i) follows from (3.10).

For (ii), we first note that since $[\mathbf{b},\mathbf{n}]\supset\mathbf{n}$, $\mathbf{n}_0\subset\mathbf{n}$ and so consists of nilpotent elements. To show that $\mathbf{n}_0$ is a subalgebra, observe that since $\mathbf{b_p}=\mathbf{a_p}\oplus\mathbf{n_p}$ and $\mathbf{a_p}$ is abelian, we may write

$$\mathbf{n}_0 = \mathbf{n_p}+[\mathbf{b_p},\mathbf{b_p}] \ .$$

Since

$$[[\mathbf{b_p},\mathbf{b_p}],\mathbf{b_p}]\subset[\mathbf{n},\mathbf{b_p}]\subset\mathbf{n_p} \ ,$$

we have

$$[[\mathbf{b_p},\mathbf{b_p}],\mathbf{n_p}]\subset\mathbf{n_p} \ ,$$

$$[[\mathbf{b_p},\mathbf{b_p}],[\mathbf{b_p},\mathbf{b_p}]]\subset[\mathbf{n_p},\mathbf{b_p}]=[\mathbf{b_p},\mathbf{b_p}]$$

so that $\mathbf{n}_0$ is a nilpotent subalgebra which is clearly τ-stable. Set $\mathbf{q}=\mathbf{n}_0^{\perp}$, then

$$\mathbf{q_p} = \mathbf{b_p} \ , \qquad \mathbf{q_k} = [\mathbf{b_p},\mathbf{b_p}]^{\perp}\cap\mathbf{k}^{\mathbf{C}} \ .$$

Thus $(\mathbf{q_k},[\mathbf{q_p},\mathbf{q_p}])$ vanishes so that

$$[\mathbf{q_k},\mathbf{q_p}]\subset\mathbf{q_p}^{\perp} = \mathbf{n_p}\subset\mathbf{q_p} \ . \qquad (8)$$

Further,

$$([\mathbf{q_k},\mathbf{q_k}],[\mathbf{q_p},\mathbf{q_p}])\subset([[\mathbf{q_k},\mathbf{q_p}],\mathbf{q_k}],\mathbf{q_p})\subset(\mathbf{n_p},\mathbf{q_p}) = 0$$

so that $[\mathbf{q_k},\mathbf{q_k}]\subset\mathbf{q_k}$. Lastly, since $[\mathbf{q_p},\mathbf{q_p}]\subset\mathbf{n}$ and $\mathbf{n}$ is isotropic, we conclude that $[\mathbf{q_p},\mathbf{q_p}]\subset\mathbf{q_k}$ and so $\mathbf{q}$ is a subalgebra which is parabolic by (4.2).

By construction

$$\mathbf{n}_0 = \mathbf{n}_0 \cap \mathbf{p}^{\mathbf{C}} + [\mathbf{n}_0 \cap \mathbf{p}^{\mathbf{C}}, \mathbf{q}_{\mathbf{p}}] ,$$

so it remains to show that $\mathbf{h}_{\mathbf{p}} \subset \mathbf{z}(\mathbf{h})$ i.e. that $\mathbf{q}_{\mathbf{p}} \cap \overline{\mathbf{q}}_{\mathbf{p}} \subset \mathbf{z}(\mathbf{q} \cap \overline{\mathbf{q}})$. But from (8), we see that $[\mathbf{q}_{\mathbf{p}}, \mathbf{q}] \subset \mathbf{n}_0$ so that

$$[\mathbf{q}_{\mathbf{p}} \cap \overline{\mathbf{q}}_{\mathbf{p}}, \mathbf{q} \cap \overline{\mathbf{q}}] \subset \mathbf{n}_0 \cap \overline{\mathbf{n}}_0 = 0$$

since $\mathbf{n}_0$ is isotropic. Thus $\mathbf{q}$ is τ-maximal and the proof is complete. □

Finally, we have the following proposition which will be useful in the sequel.

Proposition 4.30. *Let* $\mathbf{g}$ *be simple,* $\mathbf{q} \subset \mathbf{g}^{\mathbf{C}}$ *be a* τ*-maximal parabolic subalgebra and* $\mathbf{z}$ *the centre of* $\mathbf{n}$. *Then* $\mathbf{h}_{\mathbf{k}}$ *acts irreducibly on* $\mathbf{z}$ *so that either* $\mathbf{z} \subset \mathbf{k}^{\mathbf{C}}$ *or* $\mathbf{z} \subset \mathbf{p}^{\mathbf{C}}$.

Proof. By (4.3), $\mathbf{h}$ acts irreducibly on $\mathbf{z}$ and $\mathbf{h}_{\mathbf{p}} \subset \mathbf{z}(\mathbf{h})$ so that $\mathbf{h}_{\mathbf{p}}$ acts on $\mathbf{z}$ by scalars. We fix a τ-stable maximal torus $\mathbf{t} = \mathbf{t}_{\mathbf{k}} \oplus \mathbf{t}_{\mathbf{p}} \subset \mathbf{h}$, fundamental by (4.28), so that $\mathbf{h}_{\mathbf{p}} = \mathbf{t}_{\mathbf{p}}$. Now $\mathbf{q} = \mathbf{q}_I$ for some subset I of simple roots and we let θ be the corresponding highest root. Then $\mathbf{g}^{\theta} \subset \mathbf{z}$ and θ is the highest weight of the $\mathbf{h}$-representation on $\mathbf{z}$. Since $\tilde{\theta}$ is also a highest weight, we conclude that $\theta = \tilde{\theta}$ whence $[\mathbf{t}_{\mathbf{p}}, \mathbf{g}^{\theta}] = 0$.

Thus $\mathbf{h}_{\mathbf{p}}$ acts trivially on $\mathbf{z}$ so that $\mathbf{h}_{\mathbf{k}}$ acts irreducibly and the conclusion follows. □

Chapter 5. The twistor space of a Riemannian symmetric space

We now consider the twistor space $J(N)$ of a symmetric space $N = G/K$. In order to find well-behaved twistor spaces, we consider G-orbits in the locus of zeros of the Nijenhuis tensor of J_1.

In case that N is inner, we shall show that, remarkably, G acts transitively on components of the zero locus and that each orbit is a flag manifold F with $N(F) = N$. Further, these flag manifolds are embedded both J_1 and J_2 holomorphically and the canonical fibration coincides with the trace of the twistor projection. Thus we find a geometric realisation of the algebraic constructions of the previous chapter.

Notation. Throughout this chapter, N will denote an even-dimensional simply connected Riemannian symmetric G-space for some connected Lie group G. We denote the Killing form of G by B and the Maurer-Cartan form of N by β. We fix a base point $x_0 \in N$ with stabilizer K and involution τ so that the symmetric decomposition at x_0 is

$$\mathbf{g} = \mathbf{k} \oplus \mathbf{p} .$$

The metric of N restricts to an $\mathrm{Ad}(K)$-invariant inner product on $\mathbf{p}$ which we denote by $(.,.)$.

A. The zero set of the Nijenhuis tensor

Let $Z \subset J(N)$ denote the zero set of the Nijenhuis tensor of J_1. Recall from chapter 3 that the Riemannian curvature tensor, R, of N is given by

$$\beta \circ R = -\tfrac{1}{2}\mathrm{Ad}[\beta \wedge \beta]\,\beta .$$

Let $j \in J_{x_0}(N)$ with $\sqrt{-1}$-eigenspace T^+ and $\beta(T^+) = \mathbf{p}^+ \subset \mathbf{p}^{\mathbf{C}}$. Then we see from proposition (2.14) that $j \in Z$ if and only if

$$[[\mathbf{p}^+, \mathbf{p}^+], \mathbf{p}^+] \subset \mathbf{p}^+ . \tag{1}$$

We begin our study of Z by investigating its interaction with the De Rham decomposition of N. From (3.1) and (3.3), we know that N splits as a Riemannian product

$$N = N_0 \times \ldots \times N_n \tag{2}$$

where N_0 is a Euclidean space and each N_i, $i \geq 1$, is an irreducible symmetric space of semi-simple type. Corresponding to the decomposition (2), at the Lie algebra level we have a decomposition of $\mathbf{g}$ into commuting ideals compatible with the involution. Thus

$$\mathbf{g} = \mathbf{g}_0 \oplus \ldots \oplus \mathbf{g}_n$$

and setting $\mathbf{k}_i = \mathbf{k} \cap \mathbf{g}_i$, $\mathbf{p}_i = \mathbf{p} \cap \mathbf{g}_i$, we have

$$\mathbf{k} = \sum_i \mathbf{k}_i , \qquad \mathbf{p} = \sum_i \mathbf{p}_i .$$

For $i \geq 1$, $\mathbf{g}_i$ is semi-simple and we set

$$\mathbf{g}_{ss} = \sum_{i \geq 1} \mathbf{g}_i$$

and define $\mathbf{k}_{ss}$, $\mathbf{p}_{ss}$ similarly. In particular, $\mathbf{g}_{ss}$ is a semi-simple ideal of $\mathbf{g}$ which commutes with

$\mathbf{g}_0$. On the other hand, since N_0 is Euclidean, $\mathbf{p}_0$ is an abelian ideal of $\mathbf{g}$ from which we see that

$$[\mathbf{p}, \mathbf{p}] \subset \mathbf{k}_{ss} . \tag{3}$$

Finally, we may assume that the metric on $\mathbf{p}$ coincides with the metric induced by the Killing form B on $\mathbf{p}_{ss}$. (Of course, B vanishes on $\mathbf{p}_0$.)

We are now in a position to prove some algebraic consequences of formula (1).

Lemma 5.1. *Let $\mathbf{p}^+$ be a maximally isotropic subspace of $\mathbf{p}^{\mathbf{C}}$ with respect to (.,.).*

(i) $\mathbf{p}^+$ satisfies (1) if and only if $[\mathbf{p}^+, \mathbf{p}^+] \subset \mathbf{k}_{ss}^{\mathbf{C}}$ is isotropic with respect to B.

(ii) Let $\mathbf{h}_{ss} = \{\xi \in \mathbf{k}_{ss} : [\xi, \mathbf{p}^+] \subset \mathbf{p}^+\}$ and let $\mathbf{p}^+$ satisfy (1). Let $\mathbf{m}$ be the polar (with respect to B) of $\mathbf{h}_{ss}$ in $\mathbf{k}_{ss}$. Then

$$\mathbf{m}^{\mathbf{C}} = [\mathbf{p}^+, \mathbf{p}^+] \oplus \overline{[\mathbf{p}^+, \mathbf{p}^+]} .$$

In particular, with respect to B, $[\mathbf{p}^+, \mathbf{p}^+]$ is a maximal isotropic subspace of $\mathbf{m}^{\mathbf{C}}$ and the polar of $[\mathbf{p}^+, \mathbf{p}^+]$ in $\mathbf{k}_{ss}^{\mathbf{C}}$ is $\mathbf{h}_{ss}^{\mathbf{C}} \oplus [\mathbf{p}^+, \mathbf{p}^+]$.

Proof. Since $\mathbf{p}^+$ is maximal isotropic, $\mathbf{p}^+$ satisfies (1) if and only if

$$([[\mathbf{p}^+, \mathbf{p}^+], \mathbf{p}^+], \mathbf{p}^+) = 0 .$$

However, by (3), $[[\mathbf{p}^+, \mathbf{p}^+], \mathbf{p}^+] \subset \mathbf{p}_{ss}$ where (.,.) coincides with the Killing metric so that $\mathbf{p}^+$ satisfies (1) if and only if

$$B([[\mathbf{p}^+, \mathbf{p}^+], \mathbf{p}^+], \mathbf{p}^+) = 0$$

or, equivalently,

$$B([\mathbf{p}^+, \mathbf{p}^+], [\mathbf{p}^+, \mathbf{p}^+]) = 0 .$$

This establishes (i).

For (ii), we again use the maximal isotropy of $\mathbf{p}^+$ to see that, if $\xi \in \mathbf{k}_{ss}$, $\xi \in \mathbf{h}_{ss}$ if and only if $([\xi, \mathbf{p}^+], \mathbf{p}^+)$ vanishes. Since $[\xi, \mathbf{p}^+] \subset \mathbf{p}_{ss}$, this is equivalent to the vanishing of $B([\xi, \mathbf{p}^+], \mathbf{p}^+)$. Thus, by the invariance of B, $\xi \in \mathbf{h}_{ss}$ if and only if

$$B(\xi, [\mathbf{p}^+, \mathbf{p}^+]) = 0 .$$

This means that $\mathbf{h}_{ss} = \mathbf{k}_{ss} \cap [\mathbf{p}^+, \mathbf{p}^+]^{\perp}$ and the proposition now follows. □

Remark. We note that lemma (4.6) is the special case of (5.1) where $\mathbf{g}_0 = \{0\}$. Similarly, proposition (4.7) is the same special case of the following proposition.

Proposition 5.2. *Let $\mathbf{p}^+$ be a maximally isotropic subspace of $\mathbf{p}^{\mathbf{C}}$ satisfying (1) and let $\mathbf{h}_{ss} = \{\xi \in \mathbf{k}_{ss} : [\xi, \mathbf{p}^+] \subset \mathbf{p}^+\}$. Then $\mathbf{h}_{ss}$ is the centraliser of a torus in $\mathbf{k}_{ss}$.*

Proof. Let j be the Hermitian almost complex structure on $\mathbf{p}$ with $\sqrt{-1}$-eigenspace $\mathbf{p}^+$. We define a form $\omega \in \Lambda^2 \mathbf{k}^*$ by

$$\begin{aligned} \omega(\xi, \eta) &= \operatorname{trace}_{\mathbf{p}}(\operatorname{ad}\eta \circ [j, \operatorname{ad}\xi]) \\ &= \operatorname{trace}_{\mathbf{p}}(j \circ \operatorname{ad}[\xi, \eta]) . \end{aligned}$$

We claim that the radical of ω is precisely $\mathbf{h}_{ss}$. Clearly ω vanishes on $\mathbf{h}_{ss}$ since $[\operatorname{ad}\mathbf{h}_{ss}, j] = 0$. Further, if $\xi \in [\mathbf{p}^+, \mathbf{p}^+]$, from (1) we have

$$[j, \mathrm{ad}\xi]\mathbf{p}^+ = 0$$

while on $\overline{\mathbf{p}}^+$

$$[j, \mathrm{ad}\xi] = (j+\sqrt{-1})\circ\mathrm{ad}\xi \ .$$

So if $X_1, \ldots, X_r$ form a unitary basis for $\mathbf{p}^+$, we obtain

$$\begin{aligned}\omega(\xi,\bar{\xi}) &= (\mathrm{ad}\bar{\xi}[j,\mathrm{ad}\xi]X_i \, , \bar{X}_i) + (\mathrm{ad}\bar{\xi}[j,\mathrm{ad}\xi]\bar{X}_i \, , X_i) \\ &= (\mathrm{ad}\bar{\xi}[j,\mathrm{ad}\xi]\bar{X}_i \, , X_i) \\ &= -([j,\mathrm{ad}\xi]\bar{X}_i \, , \mathrm{ad}\bar{\xi}X_i)\end{aligned}$$

since $(.,.)$ is $\mathrm{ad}(\mathbf{k})$-invariant,

$$\begin{aligned}&= -((j+\sqrt{-1})[\xi,\bar{X}_i] \, , [\bar{\xi},X_i]) \\ &= \frac{-\sqrt{-1}}{2}((j+\sqrt{-1})[\xi,\bar{X}_i] \, , (j-\sqrt{-1})[\bar{\xi},X_i])\end{aligned}$$

which vanishes if and only if $[\bar{\xi},X_i]\in\mathbf{p}^+$ for each i, in which case $\xi+\bar{\xi}\in\mathbf{h}_{ss}$ by lemma (5.1ii) and so ξ vanishes. Thus ω is non-degenerate on $\mathbf{m}$ and the claim is established.
Now define $\chi\in\mathbf{k}_{ss}$ by

$$B(\chi\, , \xi) = \mathrm{trace}_{\mathbf{p}}(j\circ\mathrm{ad}\xi)$$

for $\xi\in\mathbf{k}_{ss}$. Then

$$\omega(\eta,\xi) = B(\chi\, , [\eta, \xi]) = B([\chi, \eta]\, , \xi)$$

so that the centraliser of χ is the radical of ω which we have seen to be $\mathbf{h}_{ss}$. The proposition now follows. □

We now suppose that N is inner. Then each N_i is inner and therefore even-dimensional. Further, each N_i, $i\geq 1$, is simple. We have a natural inclusion

$$J(N_0)\times\ldots\times J(N_n) \hookrightarrow J(N)$$

given by

$$(j_0,\ldots,j_n)\mapsto j_0+\ldots+j_n \ .$$

In general, an element of $J(N)$ will not split in this way but the situation is much better for Z.

Theorem 5.3. *Let N be inner and let $Z_i\subset J(N_i)$ be the zero set of the Nijenhuis tensor of J_1 on $J(N_i)$. Then*

$$Z = Z_0\times\ldots\times Z_n \ .$$

Thus any $j\in Z$ can be written as $j_0+\ldots+j_n$ with each $j_i\in Z_i$.

Proof. The assertion only concerns the fibres of Z so it suffices to prove the result at x_0. Firstly, it is clear from (1) and the fact that the $\mathbf{g}_i$ are commuting ideals of $\mathbf{g}$ that $Z_0\times\ldots\times Z_n\subset Z$. The converse is less trivial. Let $j\in Z_{x_0}$ with $\sqrt{-1}$-eigenspace $\mathbf{p}^+$. Thus $\mathbf{p}^+$ is maximally isotropic in $\mathbf{p}^{\mathbf{C}}$ and satisfies (1). We must show that

$$\mathbf{p}^+ = \sum \mathbf{p}^+\cap\mathbf{g}_i$$

and that each $\mathbf{p}^+\cap\mathbf{g}_i$ satisfies (1). We begin by splitting off the Euclidean factor. So let

$\pi_{ss}:\mathbf{g}\to\mathbf{g}_{ss}$ be projection along $\mathbf{g}_0$. We claim that $\pi_{ss}(\mathbf{p}^+)\subset\mathbf{p}^+$. For this, it suffices to prove that

$$(\pi_{ss}(\mathbf{p}^+),\mathbf{p}^+) = 0$$

or, since $\mathbf{p}_0\perp\mathbf{p}_{ss}$, that

$$(\pi_{ss}(\mathbf{p}^+),\pi_{ss}(\mathbf{p}^+)) = 0 \ .$$

Since $\pi_{ss}(\mathbf{p}^+)\subset g_{ss}$, this is equivalent to the isotropy of $\pi_{ss}(\mathbf{p}^+)$ with respect to B.

Now each N_i is simple, so $\mathbf{k}_{ss}$ has maximal rank in $\mathbf{g}_{ss}$. Thus, by (5.2), $\mathbf{h}_{ss}$ contains a maximal torus $\mathbf{t}$ of $\mathbf{g}_{ss}$. Clearly, $\pi_{ss}(\mathbf{p}^+)$ is ad($\mathbf{t}$)-invariant and so splits as a sum of root spaces. Thus, by (3.4iii), $\pi_{ss}(\mathbf{p}^+)$ is B-isotropic unless there is a root α such that $\mathbf{g}^{\alpha}$, $\mathbf{g}^{-\alpha}\subset\pi_{ss}(\mathbf{p}^+)$. But, in this case, by (3.4v), $[\mathbf{g}^{\alpha},\mathbf{g}^{-\alpha}]$ contains a non-zero element of $\mathbf{t}^{\mathbf{C}}\cap[\pi_{ss}(\mathbf{p}^+),\pi_{ss}(\mathbf{p}^+)]$. However,

$$[\pi_{ss}(\mathbf{p}^+),\pi_{ss}(\mathbf{p}^+)] = [\mathbf{p}^+,\mathbf{p}^+]$$

since $\mathbf{p}_0$ is in the centre of $\mathbf{p}$, so that $\mathbf{t}^{\mathbf{C}}\cap[\pi_{ss}(\mathbf{p}^+),\pi_{ss}(\mathbf{p}^+)]$ is zero by (5.1ii). Thus, $\pi_{ss}(\mathbf{p}^+)$ is B-isotropic and the claim is proved.

We therefore conclude that $\pi_{ss}(\mathbf{p}^+)=\mathbf{p}^+\cap\mathbf{g}_{ss}$ whence

$$\mathbf{p}^+ = \mathbf{p}^+\cap\mathbf{g}_0\oplus\mathbf{p}^+\cap\mathbf{g}_{ss} \ .$$

Further, each $\mathbf{g}_i$ is ad($\mathbf{t}$)-invariant so that any root space lies in some $\mathbf{g}_i$, $i\geq 1$, whence, since $\mathbf{p}^+\cap\mathbf{g}_{ss}$ is a sum of root spaces, we conclude that

$$\mathbf{p}^+ = \sum\mathbf{p}^+\cap\mathbf{g}_i \ .$$

Lastly, we observe that each $\mathbf{p}^+\cap\mathbf{g}_i$ satisfies (1) since $\mathbf{p}^+$ does and so the proof is complete. □

We conclude this section with two applications of these ideas to the study of local Hermitian structures on symmetric spaces.

Lemma 5.4. *Let M be an even-dimensional Riemannian manifold and J a Hermitian complex structure defined on some open subset of M. Then the local section of $J(M)$ defined by J has image in the zero set of the Nijenhuis tensor of J_1.*

Proof. A Hermitian almost complex structure is integrable if and only if its bundle of (1,0) vectors is stable under covariant differentiation by (1,0) vectors. The result now follows immediately from (2.14). □

As an immediate corollary of (5.3) and (5.4), we have

Theorem 5.5. *Let N be an inner Riemannian symmetric space. Then any Hermitian complex structure defined on an open subset of N commutes with the De Rham decomposition of N.*

In the same order of ideas is the following interesting result.

Theorem 5.6. *Let N be a Hermitian symmetric space of semi-simple type. Then any Hermitian complex structure defined on an open subset of N commutes with the invariant Hermitian symmetric complex structure.*

Proof. By (3.2), the Hermitian symmetric complex structure at $x\in N$ is given by the adjoint action of an element of the centre of the stabilizer of x, but this centre stabilises any element of Z_x by proposition (5.2) and the theorem follows. □

This result has an application to the study of the stability of harmonic maps between Hermitian symmetric spaces (see [13]).

B. G-orbits on $J(N)$

We are now in a position to see the relevance of flag manifolds and flag domains to the twistor theory of symmetric spaces:

Theorem 5.7. *Let N be a simply connected inner symmetric G-space of semi-simple type and let $Z \subset J(N)$ be the zero set of the Nijenhuis tensor of J_1. Then Z is composed of a finite number of G-orbits, each of which is a flag manifold or canonical flag domain canonically fibred over N.*

Before proving this theorem, we draw an interesting corollary to it.

Theorem 5.8. *Let N be a simply connected inner symmetric G-space and let $Z \subset J(N)$ be the zero set of the Nijenhuis tensor of J_1. Then G acts transitively on connected components of Z. In particular, each component is a submanifold of $J(N)$.*

Proof. By (5.3), it suffices to prove the theorem only for the cases that N is a Euclidean space or that N is of semi-simple type. If N is an even-dimensional Euclidean space, then $Z = J(N)$ by (2.14) since N is flat and it is well known that the Euclidean group acts transitively on the two connected components of $J(N)$. If, on the other hand, N is of semi-simple type, then each G-orbit in $J(N)$ is connected and has compact fibre whence it is closed. Thus, by (5.7), each connected component of Z consists of a finite number of closed, connected orbits and hence each must comprise a single orbit. □

Proof of (5.7). We begin by showing that any orbit in Z is equivariantly isomorphic to a canonically fibred flag manifold or canonical flag domain according as N is compact or non-compact.

Fix $j \in Z_{x_0}$ with $\sqrt{-1}$-eigenspace $\mathbf{p}^+$. From theorems (4.8) and (4.12), we see that

$$\mathbf{q} = ([\mathbf{p}^+, \mathbf{p}^+] \oplus \mathbf{p}^+)^\perp$$

is a parabolic subalgebra of $\mathbf{g}^{\mathbf{C}}$ such that the involution at x_0 is canonical for $\mathbf{q}$. Further, if $\mathbf{g}$ is non-compact, we see that $(\mathbf{g}, \mathbf{q})$ is a canonical pair.

Let $P \subset G^{\mathbf{C}}$ be the parabolic subgroup of $G^{\mathbf{C}}$ with Lie algebra $\mathbf{q}$. Then the stabiliser of j is $K \cap P$. However, $G \cap P$ is connected (by (4.13), if G is non-compact) and $\mathbf{k} \cap \mathbf{p} = \mathbf{g} \cap \mathbf{p}$ so that we conclude that $K \cap P = G \cap P$. Thus, the G-orbit of j is equivariantly isomorphic to the G-conjugacy class of $\mathbf{q}$ which is a flag manifold or canonical flag domain that we denote by X. Further, the projection $\pi : J(N) \to N$ is G-equivariant so that the restriction of π to $G \cdot j$ is a homogeneous fibration. Moreover, since the involution at x_0 is canonical for $\mathbf{q}$, we deduce that $N = N(X)$ and that π is a canonical fibration.

Thus each G-orbit in Z gives rise to a unique G-conjugacy class of parabolic subalgebras of $\mathbf{g}^{\mathbf{C}}$. We now consider how many times each such conjugacy class can arise. So suppose that $j_1, j_2 \in Z_{x_0}$ generate G-conjugate parabolic subalgebras $\mathbf{q}_1, \mathbf{q}_2$. Since the canonical involutions coincide and K is connected, we conclude that the $\mathbf{q}_i$ are conjugate via an element of $N_G(K)$

while the j_i belong to the same orbit if and only if the $\mathbf{q}_i$ are K-conjugate. Thus each G-conjugacy class that appears in Z appears $|\Sigma_N|$ times. Finally, by [80, theorem 2.6], there are but a finite number of G-conjugacy classes of parabolic subalgebras of $\mathbf{g}^{\mathbf{C}}$ and so we conclude that Z is a finite union of G-orbits. □

Remarks. (i) We have seen that if N is of compact type, each flag manifold that appears in Z appears $|\Sigma_N|$ times. On the other hand, if N is of non-compact type, Σ_N is trivial and so a given canonical flag domain appears at most once in Z. However, it is not difficult to show that if a flag domain contained in a flag manifold F does appear in Z, then $|\Sigma_{\tilde{N}}|$ distinct canonical flag domains of F appear in Z, where $\tilde{N}$ is the compact dual of N. Conversely, in this case, F will appear in $Z \subset J(\tilde{N})$. Thus we conclude that the number of components of Z is the same for both N and $\tilde{N}$.

(ii) If N is even-dimensional but non-inner, the situation is much less well behaved. In this case, Z is non-empty but in general G does not act transitively on connected components of Z. In fact, one can show that, for N non-inner, each component of Z arises as the bundle of Hermitian almost complex structures on a certain torus bundle over a G-orbit in Z. We shall not pursue this topic further.

C. Twistor fibrations of flag manifolds

Theorem (5.7) can be considerably strengthened: any flag manifold or flag domain X can be embedded as an orbit in $Z \subset J(N(X))$ in a G-equivariant way $|\Sigma_{N(X)}|$ times and, further, these embeddings intertwine the J_1, J_2 almost complex structures on X and $J(N(X))$.

For this we argue as follows: fix a canonical fibration $\pi: X \to N(X)$ and, for $x \in X$, set $j(x)$ to be the almost Hermitian structure on $T_{\pi(x)}N(X)$ with $\sqrt{-1}$-eigenspace $[\mathbf{p}^{\mathbf{C}} \cap \mathbf{n}]_x$. Clearly, $j: X \to J(N(X))$ is a G-equivariant fibre map which is in fact an embedding since the normaliser of $\mathbf{p}^{\mathbf{C}} \cap \mathbf{n}$ is precisely $G \cap P$. Also, we see that j is precisely the realisation of X as the orbit in $J(N(X))$ of $j(x)$ as in theorem (5.7). We have now

Proposition 5.9 *The image of j is contained in $Z \subset J(N(X))$ and further, for $i = 1,2$,*

$$J_i \circ dj = dj \circ J_i \ .$$

Proof. That $\mathrm{Im}\, j \subset Z$ is immediate from formula (1) and the isotropy of $\mathbf{n}$.

For the holomorphicity of j, set $N(X) = N$ and let the Maurer-Cartan forms of X and N be β^X, β^N respectively. From (1.8) we have

$$\pi^* \beta^N = \pi_{\mathbf{p}} \circ \beta^X \tag{4}$$

where $\pi_{\mathbf{p}}: \underline{\mathbf{g}} \to [\mathbf{p}]$ is orthogonal projection. Thus if ∇^N, D^X are the Levi-Civita connection on N and the canonical connection on X respectively, we see from proposition (1.1) and lemma (1.6) that

$$\pi^{-1} \nabla^N = D^X + \mathrm{ad}(1 - \pi_{\mathbf{p}}) \beta^X \ . \tag{5}$$

Now, the bundle $\underline{j}$ of (1,0) vectors on $\pi^{-1}TN$ determined by j is given by

$$\pi^{-1} \beta^N(\underline{j}) = [\mathbf{p} \cap \mathbf{n}] \ .$$

From (4) we see that

$$\pi_*(T^{1,0}_{J_i} X) \subset \underline{j}$$

while from (5) we conclude that $\pi^{-1}\nabla_Z^N$ will preserve $C^\infty([\mathbf{p}\cap\mathbf{n}])$ if and only if $\mathrm{ad}(1-\pi_{\mathbf{p}})\beta^X(Z)$ does since $[\mathbf{p}\cap\mathbf{n}]$ is D^X-parallel.

Now

$$(1-\pi_{\mathbf{p}})\beta^X(T_{J_1}^{1,0}X) = (1-\pi_{\mathbf{p}})\beta^X(T_{J_2}^{0,1}X) = [\mathbf{k}\cap\mathbf{n}]$$

and of course $[\mathbf{k}\cap\mathbf{n}, \mathbf{p}\cap\mathbf{n}]\subset\mathbf{p}\cap\mathbf{n}$ so that we have

$$\pi^{-1}\nabla_Z^N C^\infty(\underline{j})\subset C^\infty(\underline{j})$$

for both $Z\in T_{J_1}^{1,0}X$ and $Z\in T_{J_2}^{0,1}X$. An application of theorem (2.3) completes the proof.

□

Corollary 5.10. *The canonical fibrations are twistor fibrations of (X,J_2) over $N(X)$.*

As we vary the canonical fibrations we obtain $|\Sigma_N|$ such embeddings and from theorem (5.7) we see that these are precisely the realisations of the X as the connected components of $Z\subset J(N(X))$.

To sum up: any canonical fibration of a flag manifold X may be realised as the trace of the projection $J(N(X))\rightarrow N(X)$ on an orbit in Z and then the J_i structures on X are just the trace of those on $J(N(X))$. This gives a geometrical interpretation of the essentially algebraic constructions of the previous section.

Examples and further results

Let us consider the case that N is an even-dimensional sphere S^{2n}. Since N is conformally flat, $Z=J(N)$ and so has two components corresponding to the two possible orientations. Further, $\Sigma_N = \mathbf{Z}_2$ in this case so that there is a unique flag manifold realising the components of Z with canonical fibrations permuted by the antipodal map. We deduce from (4.18) and (4.19) that, for $n\geq 2$, this flag manifold has height two and conclude:

Proposition 5.11. *If F is a flag manifold and $N(F)=S^{2n}$, then F has height not exceeding two.*

In fact, one can check by direct calculation that $F=\mathbf{SO}(2n+1)/\mathbf{U}(n)$.

Theorems (5.7) and (5.8) provide a powerful tool for the study of Hermitian complex structures on Riemannian symmetric spaces. Indeed, D. Burns[1], using a similar analysis of $Z\subset J(\mathbf{C}P^n)$ together with results from Algebraic Geometry, has shown that any globally defined Hermitian complex structure on $\mathbf{C}P^n$ coincides with the standard one up to sign. This leads us to make the following

Conjecture. *If N is Hermitian symmetric of semi-simple type, then any globally defined Hermitian complex structure on N is Hermitian symmetric.*

Finally, we note that our methods provide new proofs of some related results of Bryant [11]. We sketch this development below.

Bryant considered orbits of $J(N)$ on which J_1 is integrable and the horizontal (1,0) vectors forms a holomorphic distribution. If $\tilde{Z}$ denotes the zero-set of the corresponding integrability

---

[1] Private communication.

conditions on $J(N)$, he was able to show that G acts transitively on components of $\tilde{Z}$ and to classify the flag manifolds that arise. In fact, $\tilde{Z}$ is the set of $j \in J(N)$ whose $\sqrt{-1}$-eigenspaces satisfy

$$R(T^+, T^+)\, T^+ = 0 \, .$$

In our context, this means considering $\mathbf{p}^+ \subset \mathbf{p}^{\mathbb{C}}$ such that

$$[[\mathbf{p}^+, \mathbf{p}^+], \mathbf{p}^+] = 0 \, . \tag{6}$$

As an immediate consequence of (6), we see that $[\mathbf{p}^+, \mathbf{p}^+]$ is abelian while, using (5.1ii), we conclude that

$$[\mathbf{p}^+, \overline{\mathbf{p}}^+] \subset \mathbf{h}_{ss} \, .$$

A direct calculation now shows that $\mathbf{p}^+$ is isotropic with respect to the Killing form of $\mathbf{g}$ and we may conclude that

$$\mathbf{p}^+ = \mathbf{p}^+ \cap \mathbf{g}_0 \oplus \mathbf{p}^+ \cap \mathbf{g}_{ss}$$

with both summands satisfying (6). Specialising now to the case of semi-simple $\mathbf{g}$, we see from (6) that $[\mathbf{p}^+, \mathbf{p}^+] \oplus \mathbf{p}^+$ is two-step nilpotent and we can then show that

$$\mathbf{q} = ([\mathbf{p}^+, \mathbf{p}^+] \oplus \mathbf{p}^+)^\perp$$

is a parabolic subalgebra without the *a priori* assumption that the involution be inner.

In this way, we can recover by algebraic means the results of Bryant that

(a) $\tilde{Z}$ splits as a direct product according to the De Rham decomposition of N;
(b) $\tilde{Z} = \varnothing$ unless N is inner;
(c) G is transitive on connected components of $\tilde{Z}$.

It is also clear that, for N of semi-simple type, $X \subset \tilde{Z}$ if and only if X has height not exceeding two.

Chapter 6. Twistor Lifts over Riemannian Symmetric Spaces

We saw in Chapter 2 a number of general results which exhibited conformal harmonic maps of a Riemann surface M into an even dimensional manifold N as projections of J_2-holomorphic curves in $J(N)$. In Chapter 5, we found that when N is an inner symmetric space the zero set of the Nijenhuis tensor of J_1 on $J(N)$ consisted of a finite number of flag manifolds or flag domains canonically fibred over N. We can therefore ask if there is a J_2-holomorphic lift whose image lies in such a homogeneous twistor space. In this chapter, we shall convert this into an algebraic problem and describe some solutions. In particular, we shall obtain a unified proof of some results of Calabi [24], Eells-Wood [34] and Bryant [10] concerning minimal immersions of 2-spheres in even-dimensional spheres and complex projective spaces.

Remark. Since the Riemannian symmetric spaces of non-compact type are non-positively curved and simply-connected, they are convex-supporting and so are not the target of any non-constant harmonic map of a compact manifold [37]. Thus we shall restrict attention in this chapter and the next to symmetric spaces of compact type.

A. Twistor lifts and bundles of parabolic subalgebras

Let N be an inner symmetric space of compact type with metric induced from the Killing form and symmetric decomposition $\mathbf{g}=\mathbf{k}\oplus\mathbf{p}$. Let $\varphi:M\to N$ be a harmonic map of a Riemann surface into N. We wish to find a flag manifold F canonically fibred over N and a J_2-holomorphic map $\psi:M\to F$ so that ψ covers φ. We shall call such a map a *twistor lift* of φ.

Let $Z\subset J(N)$ be the zero set of the Nijenhuis tensor of J_1. Recall from chapter 5 that the connected components of Z are precisely the flag manifolds which fibre canonically over N. Thus, if M is connected, it suffices to find a lift into $J(N)$ with image in Z. For this, we recall from (5.1) that if $j\in J(N)$ has (1,0)-space $\mathbf{p}^+\subset\mathbf{p}^{\mathbf{C}}$, then j lies in Z if and only if $[\mathbf{p}^+,\mathbf{p}^+]$ is isotropic. To summarise:

Theorem 6.1. *Let $\varphi:M\to N$ be a smooth map of a connected Riemannian manifold into N. Let $T^+\subset\varphi^{-1}[\mathbf{p}^{\mathbf{C}}]$ be a maximally isotropic sub-bundle. Then the map $\psi:M\to J(N)$ given by*

$$\underline{\psi}=T^+$$

has image in a component of Z if and only if $[T^+,T^+]\subset\varphi^{-1}[\mathbf{k}^{\mathbf{C}}]$ is isotropic.

Let us specialise to the case where M is a Riemann surface and consider when a map $\psi:M\to F\subset Z$ is J_2-holomorphic. By (5.9), this is the case if and only if ψ is J_2-holomorphic as a map into $J(N)$ and so we may apply (2.3) to conclude:

Theorem 6.2. *Let $\varphi:M\to N$ be a map of a connected Riemann surface into N. Then there is a 1–1 correspondence between J_2-holomorphic maps $\psi:M\to J(N)$ covering φ which have image in a component of Z and maximally isotropic sub-bundles T^+ of $\varphi^{-1}[\mathbf{p}^{\mathbf{C}}]$ satisfying*

(i) $[T^+,T^+]$ is isotropic;

(ii) T^+ is holomorphic with respect to the Koszul-Malgrange holomorphic structure induced by

the pull-back of the Levi-Civita connection of N;

(iii) $\varphi_*(T^{1,0}M) \subset T^+$.

The correspondence is given by

$$\underline{\psi} = T^+ .$$

Thus to find a twistor lift into a canonically fibred flag manifold, it suffices to find a maximally isotropic holomorphic sub-bundle of $\varphi^{-1}[\mathbf{p}^{\mathbf{C}}]$ which contains $\varphi_*(T^{1,0}M)$ and has isotropic bracket.

We may gain a rather more intrinsic viewpoint on these questions by considering sub-bundles of parabolic subalgebras in $\underline{\mathbf{g}}^{\mathbf{C}}$. Let F be a flag manifold with (complex) isotropy bundle $[\mathbf{q}]$. Then a map $\psi: M \to F$ gives rise to a sub-bundle of parabolic subalgebras $\psi^{-1}[\mathbf{q}] \subset \underline{\mathbf{g}}^{\mathbf{C}}$. Conversely, since there are only a finite number of conjugacy classes of parabolic subalgebras, all fibres of a bundle of such are conjugate. Thus any sub-bundle of parabolic subalgebras defines a map into a flag manifold and is the pull-back by that map of the isotropy bundle of the flag manifold.

Suppose now that F fibres canonically over N and $\varphi: M \to N$ is the projection of $\psi: M \to F$. We obtain a fibre-wise symmetric decomposition

$$\underline{\mathbf{g}}^{\mathbf{C}} = \varphi^{-1}[\mathbf{k}^{\mathbf{C}}] \oplus \varphi^{-1}[\mathbf{p}^{\mathbf{C}}] \tag{1}$$

and we see that this decomposition is fibre-wise canonical for $\psi^{-1}[\mathbf{q}]$. Thus we conclude that a map into a canonically fibred flag manifold which covers φ is equivalent to a bundle of parabolic subalgebras in $\underline{\mathbf{g}}^{\mathbf{C}}$ for which the splitting (1) is canonical.

Let us now relate this to the set-up of (6.1). First note that if $\psi: M \to Z$ defines a maximally isotropic sub-bundle $T^+ \subset \varphi^{-1}[\mathbf{p}^{\mathbf{C}}]$ then as in (5.7) we obtain a bundle of parabolic subalgebras by

$$\psi^{-1}[\mathbf{q}] = (T^+ \oplus [T^+, T^+])^{\perp} , \tag{2}$$

while, given $\psi: M \to F$, we obtain T^+ by

$$T^+ = \varphi^{-1}[\mathbf{p}^{\mathbf{C}}] \cap \psi^{-1}[\mathbf{q}] . \tag{3}$$

Finally, we suppose that M is a Riemann surface and consider the holomorphicity of ψ. Once and for all, we equip $\underline{\mathbf{g}}^{\mathbf{C}}$ with the pull-back by φ of the canonical connection on N and with the corresponding Koszul-Malgrange holomorphic structure. Then the splitting (1) is parallel with the connection coinciding with the pull-back of the Levi-Civita connection on the second summand. Further, the Killing form and the adjoint representation are parallel whence, from (2) and (3), we conclude that $\psi^{-1}[\mathbf{q}]$ is a holomorphic sub-bundle of $\underline{\mathbf{g}}^{\mathbf{C}}$ if and only if T^+ is a holomorphic sub-bundle of $\varphi^{-1}[\mathbf{p}^{\mathbf{C}}]$. Thus we may restate (6.2) as follows:

Theorem 6.3. *Let $\varphi: M \to N$ be a map of a connected Riemann surface into N. Then there is a 1–1 correspondence between J_2-holomorphic maps ψ of M into a flag manifold canonically fibred over N which cover φ and holomorphic sub-bundles of parabolic subalgebras* $\mathbf{Q}$ *of* $\underline{\mathbf{g}}^{\mathbf{C}}$ *satisfying*

(i) the symmetric decomposition (1) is fibre-wise canonical for $\mathbf{Q}$*;*

(ii) $\varphi_*(T^{1,0}M) \subset \mathbf{Q} \cap \varphi^{-1}\mathbf{p}^{\mathbf{C}}$.

This correspondence is given by

$$Q = \psi^{-1}[q] \ .$$

Both these pictures of twistor lifts will be of use in the sequel.

B. Twistor lifts in the Hermitian symmetric case

In chapter 2, we saw that a sufficiently ramified harmonic map of a closed Riemann surface M into a Kähler manifold N was covered by a J_2-holomorphic map $M \to J(N)$ (see (2.10) and (2.11)). We will now show that in case N is locally Hermitian symmetric these lifts have image in $Z \subset J(N)$. Thus, in particular, if M is connected and N is a Hermitian symmetric space of compact type, a sufficiently ramified harmonic map $M \to N$ has a twistor lift into a flag manifold canonically fibred over N.

Let us recall the situation of Chapter 2. Let $\varphi: M \to N$ be a harmonic map of a closed Riemann surface into a Kähler manifold N. We set $T = \varphi^{-1}TN^{\mathbb{C}}$ and $D = \varphi^{-1}\nabla^N$, the pull-back of the Levi-Civita connection on N. We then have the D-parallel type decomposition

$$T = T' \oplus T'' \ . \tag{4}$$

Giving T the Koszul-Malgrange structure induced by D, we get the HN filtration of T,

$$\ldots T_1 \subset T_0 \subset \ldots \subset T$$

compatible with (4),

$$T_i = T'_i \oplus T''_i \ ,$$

with each T_i/T_{i+1}, T'_i/T'_{i+1} etc, either zero or semi-stable of slope i.

Now, by (2.9), for each i, $T'_i \oplus T''_{1-i}$ is a maximally isotropic holomorphic sub-bundle of T and so defines a map $\psi_i: M \to J(N)$ which will be J_2-holomorphic if φ is sufficiently ramified to force $\varphi_*(T^{1,0}M)$ to live in $\underline{\psi_i} = T'_i \oplus T''_{1-i}$. We now show that if N is locally symmetric then each ψ_i has image in Z.

Lemma 6.4. *If N is a locally Hermitian symmetric space, then the maps $\psi_i: M \to J(N)$ have image in Z.*

Proof. N is locally symmetric if and only if the curvature tensor R^N is ∇^N-parallel. In this case $\varphi^{-1}R^N$ is D-parallel and hence holomorphic. By (2.10), it suffices to prove that

$$\varphi^{-1}R^N(\underline{\psi_i}, \underline{\psi_i})\underline{\psi_i} \subset \underline{\psi_i} \ . \tag{5}$$

For this, first observe that since N is Kähler, R^N is a $(1,1)$ form with values in the type-preserving endomorphisms. Thus (5) is equivalent to

$$\varphi^{-1}R^N(T'_i, T''_{1-i})T'_i \subset T'_i \tag{6a}$$

$$\varphi^{-1}R^N(T'_i, T''_{1-i})T''_{1-i} \subset T''_{1-i} \ . \tag{6b}$$

However, by (2.7), $\inf(T'_i \otimes T''_{1-i} \otimes T'_i) \geq i+1$ while $\sup T'/T'_i < i$ and this establishes (6a). An identical argument gives (6b) and we are done. □

We may now specialise theorems (2.10) and (2.11) to our situation.

Theorem 6.5. *Let M_g be a closed Riemann surface of genus g and $\varphi: M_g \to N$ a harmonic map into a locally Hermitian symmetric space. If $r_\varphi > 2g-2$ then ψ_0 and ψ_1 are J_2-holomorphic* *lifts with image in Z.*

Theorem 6.6. *If the holomorphic and antiholomorphic ramifications r_φ', r_φ'' of a non $\pm$holomorphic harmonic map $\varphi:M_g \to N$ of a Riemann surface of genus g into a locally Hermitian symmetric space satisfy*

$$r_\varphi' + r_\varphi'' \geq 4g-3$$

then, for all k with $2g-1-r_\varphi'' \leq k \leq r_\varphi' - 2g-2$, ψ_k is a J_2-holomorphic lift with image in Z.

It is instructive to see these results from the view-point of bundles of parabolic subalgebras. Consider a holomorphic bundle E of complex Lie algebras with HN filtration

$$\ldots E_1 \subset E_0 \subset \ldots \subset E\ .$$

Since the bracket is holomorphic, we conclude from (2.7) that

$$[E_i, E_j] \subset E_{i+j}\ .$$

In particular, E_0 is a bundle of subalgebras in which E_1 is a bundle of nilpotent ideals. If we further assume that E is a bundle of semi-simple Lie algebras then the Killing form is holomorphic and, by (2.8), E_1 is the polar of E_0. We may now apply (4.2) to obtain the following result of Atiyah-Bott [2].

Lemma 6.7. *If E is a holomorphic bundle of complex semi-simple Lie algebras then E_0 is a bundle of parabolic subalgebras with nilradical bundle E_1.*

Let $\varphi:M \to N$ be a map of a closed Riemann surface into a symmetric space of semi-simple type. We would like to apply (6.7) to the bundle $E = \underline{\mathbf{g}}^{\mathbf{C}}$ over M with its Koszul-Malgrange holomorphic structure. Certainly we obtain a bundle of parabolic subalgebras compatible with the symmetric decomposition (1) and if φ is harmonic and sufficiently ramified then $\varphi_*(T^{1,0}M) \subset E_1$. However, there is no guarantee that (1) is canonical for E_0 i.e. that $E_0 \cap \varphi^{-1}[\mathbf{p}^{\mathbf{C}}]$ is maximal isotropic in $\varphi^{-1}[\mathbf{p}^{\mathbf{C}}]$ and generates E_1. Thus E_0 need not define a twistor lift for φ and some modification of the bundle may be required.

As we have seen above, this difficulty can be resolved in case that N is Hermitian symmetric. Indeed, T' and T'' are abelian subalgebras of E so that by (2.7)

$$[T'_i \oplus T''_{1-i}, T'_i \oplus T''_{1-i}] = [T'_i, T''_{1-i}] \subset E_1 \cap \varphi^{-1}[\mathbf{k}^{\mathbf{C}}]$$

whence $[T'_i \oplus T''_{1-i}, T'_i \oplus T''_{1-i}]$ is isotropic. Thus each ψ_i defines a bundle of parabolic subalgebras for which (1) is canonical by the usual prescription

$$\mathbf{Q} = ([\underline{\psi_i}, \underline{\psi_i}] \oplus \underline{\psi_i})^\perp\ .$$

These are of course the twistor lifts provided by theorems (6.5) and (6.6).

C. The Birkhoff-Grothendieck Theorem

The twistor lifts we have just constructed were built by modifying the parabolic bundle E_0 and our method required that N be Hermitian symmetric. To construct twistor lifts over other symmetric spaces, we need another technique for finding suitable maximally isotropic sub-bundles of $\varphi^{-1}[\mathbf{p}^{\mathbf{C}}]$. In case that $M = S^2$, just as in chapter 2, such a technique is available using Grothendieck's refinement of the Harder-Narasimhan theory. We require Grothendieck's generalisation [40] of the splitting theorem (2.12) to arbitrary holomorphic principal bundles over

S^2 with reductive structure group, which we now recall.

Let $P \to S^2$ be a holomorphic principal $G^{\mathbf{C}}$ bundle over S^2 with $G^{\mathbf{C}}$ a complex reductive group. An example is the dual of the Hopf bundle with its zero section deleted, $Y \to S^2$ which is a principal $\mathbf{C}^*$ bundle of degree 1. Grothendieck's Theorem says this is essentially the only holomorphic principal bundle over S^2 in the sense that all other bundles are associated to it:

Theorem 6.8 *If $P \to S^2$ is a holomorphic principal $G^{\mathbf{C}}$ bundle with $G^{\mathbf{C}}$ a complex reductive Lie group then there is a holomorphic homomorphism $\chi: \mathbf{C}^* \to G^{\mathbf{C}}$ and an isomorphism $P \to Y \times_{\mathbf{C}^*} G^{\mathbf{C}}$ of holomorphic principal $G^{\mathbf{C}}$ bundles. χ is unique up to conjugacy in $G^{\mathbf{C}}$.*

We want to reinterpret this in terms of the associated bundles of Lie algebras $P[\mathbf{g}^{\mathbf{C}}]$. Any homomorphism $\chi: \mathbf{C}^* \to G^{\mathbf{C}}$ has the form

$$\chi(e^{iz}) = \exp_G z\xi \tag{7}$$

for some $\xi \in \mathbf{g}^{\mathbf{C}}$. By the theorem ξ is determined up to conjugacy and by (7) it is in the unit lattice of $G^{\mathbf{C}}$: $\exp_G 2\pi\xi = e$. It follows ξ is semisimple, and its conjugacy class meets a compact real form $\mathbf{g}$ of $\mathbf{g}^{\mathbf{C}}$. If $\mathbf{g}$ is given, we shall assume χ has been chosen in its conjugacy class so that $\xi \in \mathbf{g}$. ξ is then unique up to conjugacy in G.

Let $\mathbf{t}$ be a maximal torus of $\mathbf{g}$ containing ξ and $\mathbf{t}^{\mathbf{C}}$ be its complexification. Then $\mathbf{t}^{\mathbf{C}}$ is a Cartan subalgebra of $\mathbf{g}^{\mathbf{C}}$. Let T and $T^{\mathbf{C}}$ be the corresponding subgroups of G and $G^{\mathbf{C}}$, so χ factors through $T^{\mathbf{C}}$. We can then form the bundle $\mathbf{a}$ given by

$$\mathbf{a} = Y \times_{\mathbf{C}^*} \mathbf{t}^{\mathbf{C}} \subset Y \times_{\mathbf{C}^*} \mathbf{g}^{\mathbf{C}} = P[\mathbf{g}^{\mathbf{C}}] .$$

Since $\mathbf{t}^{\mathbf{C}}$ is abelian $\mathbf{a}$ is a holomorphically trivial sub-bundle of $P[\mathbf{g}^{\mathbf{C}}]$ and has a section corresponding with $\xi \in \mathbf{t}^{\mathbf{C}}$ which we denote by the same symbol. Each fibre of $\mathbf{a}$ is a Cartan subalgebra of the corresponding fibre of $P[\mathbf{g}^{\mathbf{C}}]$ so we can take the corresponding root space decomposition. These fit together to give root space line bundles $\mathbf{g}^\alpha \subset P[\mathbf{g}^{\mathbf{C}}]$ and

$$P[\mathbf{g}^{\mathbf{C}}] = \mathbf{a} \oplus \sum \mathbf{g}^\alpha$$

is a direct sum. Here the sum is over the set Δ of root sections so Δ is a subset of the space $H^0(\mathbf{a}^*)$ of holomorphic sections of $\mathbf{a}^*$. Since $\mathbf{a}$ and hence $\mathbf{a}^*$ are holomorphically trivial, $H^0(\mathbf{a}^*)$ is l-dimensional where l is the rank of $G^{\mathbf{C}}$, and Δ is a copy of the root system $\Delta(\mathbf{g}^{\mathbf{C}}, \mathbf{t}^{\mathbf{C}})$.

Proposition 6.9. *For each $\alpha \in \Delta$, $\alpha(\xi)/\sqrt{-1}$ is an integer which gives the degree of the line bundle $\mathbf{g}^\alpha$.*

Proof. $\mathbf{g}^\alpha$ is associated to Y via the composition $\mathbf{C}^* \to T^{\mathbf{C}} \to \mathbf{C}^*$ given at the Lie algebra level by $\sqrt{-1}z \mapsto z\xi \mapsto \alpha(z\xi)$. On $\mathbf{C}^*$ the associating map is thus $z \mapsto z^{\alpha(\xi)/\sqrt{-1}}$ and the conclusion follows. □

A similar argument shows that if V is any holomorphic representation of $G^{\mathbf{C}}$ and $V = \sum V_\mu$ is its decomposition into weight spaces under $T^{\mathbf{C}}$ then $P[V] = \sum Y[V_\mu]$ is a holomorphic direct sum and each $Y[V_\mu]$ is a direct sum of line bundles of the same degree $\mu(\xi)/\sqrt{-1}$. Thus each $Y[V_\mu]$ is semistable of slope $\mu(\xi)/\sqrt{-1}$

D. Bundles of parabolic subalgebras and harmonic maps

In this section we take $N = G/K$ to be an arbitrary symmetric space of compact type, not necessarily inner or even-dimensional. This generality is required for an application of the structure theory to the stability of harmonic maps of 2-spheres into symmetric spaces which we shall make in the next chapter.

Suppose we have a smooth map $\varphi: S^2 \to G/K$. Then we get a principal K-bundle $\varphi^{-1}G$ with connection over S^2 by pulling back the bundle $G \to G/K$. Complexifying and taking the Koszul-Malgrange structure, we get a holomorphic principal $K^{\mathbf{C}}$ bundle. Both $\varphi^{-1}[\mathbf{k}^{\mathbf{C}}]$ and $\varphi^{-1}[\mathbf{g}^{\mathbf{C}}]$ are holomorphic bundles of Lie algebras associated to this principal bundle. If we take a maximal torus $\mathbf{t_k}$ for $\mathbf{k}$ then applying theorem (6.8) we get a holomorphically trivial Cartan subalgebra bundle $\mathbf{a_k}$ of $\varphi^{-1}[\mathbf{k}^{\mathbf{C}}]$ and a holomorphic section ξ of $\mathbf{a_k}$, together with a set of roots $\Delta_{\mathbf{k}} \subset H^0(\mathbf{a_k}^*)$ and a decomposition

$$\varphi^{-1}[\mathbf{k}^{\mathbf{C}}] = \mathbf{a_k} \oplus \sum \mathbf{k}^{\alpha} .$$

Each $\mathbf{k}^{\alpha}$ is a line bundle of degree $\alpha(\xi)/\sqrt{-1}$.

$\mathbf{p}$ is a representation of K and so complexifies to give a holomorphic representation of $K^{\mathbf{C}}$. If $\mathbf{t_p}$ is the centralizer of $\mathbf{t_k}$ in $\mathbf{p}$ we have the associated zero weight bundle $\mathbf{a_p} \subset \varphi^{-1}[\mathbf{p}^{\mathbf{C}}]$ and $\mathbf{a} = \mathbf{a_k} \oplus \mathbf{a_p}$ is a holomorphically trivial bundle of fundamental Cartan subalgebras of $\underline{\mathbf{g}}^{\mathbf{C}}$ by (4.29i). Then we have a holomorphic root bundle decomposition

$$\underline{\mathbf{g}}^{\mathbf{C}} = \mathbf{a} + \sum \mathbf{g}^{\alpha}$$

which is a sum over a set Δ of root sections and each $\mathbf{g}^{\alpha}$ has degree $\alpha(\xi)/\sqrt{-1}$.

Let τ be the section of involutions with fixed set $\varphi^{-1}[\mathbf{k}^{\mathbf{C}}]$. Then clearly $\mathbf{a}$ and hence Δ are τ-stable. Let us choose a τ-stable positive root system such that $\alpha(\xi)/\sqrt{-1} \geq 0$ for each positive root α and let $\mathbf{b}$ be the corresponding holomorphic bundle of Borel subalgebras with nilradical bundle $\mathbf{n}$. Recall from (4.29ii) that if $\mathbf{n_p} = \mathbf{n} \cap \varphi^{-1}[\mathbf{p}^{\mathbf{C}}]$, $\mathbf{b_p} = \mathbf{b} \cap \varphi^{-1}[\mathbf{p}^{\mathbf{C}}]$, then

$$\mathbf{Q} = (\mathbf{n_p} \oplus [\mathbf{n_p}, \mathbf{b_p}])^{\perp}$$

defines a bundle of τ-maximal parabolic subalgebras which is clearly holomorphic . Further, since $\mathbf{b} \subset E_0$, we have $E_1 \subset \mathbf{n}$ whence, if φ is harmonic, $\varphi_*(T^{1,0}S^2) \subset E_2 \subset \mathbf{n}$. Further $\mathbf{Q} \cap \varphi^{-1}[\mathbf{p}^{\mathbf{C}}] = \mathbf{n_p} \oplus \mathbf{a_p}$ and so is contained in E_0 while $E_1 \subset \mathbf{n} \subset \mathbf{b} \subset \mathbf{Q}$. Thus we have shown the main theorem of this section:

Theorem 6.10. *Let $\varphi: S^2 \to N$ be a harmonic map into a symmetric space of compact type. Then there is a holomorphic bundle $\mathbf{Q} \subset \underline{\mathbf{g}}^{\mathbf{C}}$ of τ-maximal parabolic subalgebras with nilradical bundle $\mathbf{N}$ satisfying*

(i) $E_1 \subset \mathbf{Q}$;

(ii) $E_1 \cap \varphi^{-1}[\mathbf{p}^{\mathbf{C}}] \subset \mathbf{N} \cap \varphi^{-1}[\mathbf{p}^{\mathbf{C}}] \subset \mathbf{Q} \cap \varphi^{-1}[\mathbf{p}^{\mathbf{C}}] \subset E_0 \cap \varphi^{-1}[\mathbf{p}^{\mathbf{C}}]$;

(iii) $\varphi_*(T^{1,0}S^2) \subset \mathbf{N}$.

In the next chapter, we shall see an application of this theorem to the stability of harmonic 2-spheres in symmetric spaces but let us now use it to find twistor lifts in case N is inner. For this, recall that if τ is inner, $\mathbf{Q}$ is τ-maximal if and only if τ is canonical for $\mathbf{Q}$. Thus in this case, $\mathbf{Q}$ defines a twistor lift of φ:

Theorem 6.11. *Let $\varphi: S^2 \to N$ be a harmonic map into a compact inner symmetric space. Then* $\mathbf{Q}$ *defines a J_2-holomorphic lift ψ of φ into a flag manifold canonically fibred over N. Moreover*

ψ satisfies

$$E_1 \cap \varphi^{-1}[\mathbf{p}^{\mathbf{C}}] \subset \underline{\psi} \subset E_0 \cap \varphi^{-1}[\mathbf{p}^{\mathbf{C}}] .$$

Remark. If N is even-dimensional but non-inner, we may still prove an analogue of (6.11) as follows. Since N is even-dimensional, it follows that $\mathbf{t_p}$ is even-dimensional and since $\mathbf{a_p}$ is holomorphically trivial, we can find a maximally isotropic holomorphic sub-bundle $\mathbf{a_p^+} \subset \mathbf{a_p}$. Then setting

$$\underline{\psi} = \mathbf{n_p} \oplus \mathbf{a_p^+}$$

defines a maximally isotropic sub-bundle of $\varphi^{-1}[\mathbf{p}^{\mathbf{C}}]$ whose bracket is isotropic being contained in the nilradical of $\mathbf{Q}$. Thus $\psi : S^2 \to J(N)$ is a J_2-holomorphic map with image in Z. However, since we do not possess a satisfactory structure theory for Z in this case, we shall not pursue this matter further.

E. Holomorphic differentials.

If $\varphi : M^2 \to (N,h)$ is a harmonic map of a Riemann surface into any Riemannian manifold then $(\varphi^* h)^{2,0}$ is a holomorphic quadratic differential. If $M = S^2$ there are no such differentials and so $\varphi^* h$ is of type (1,1), or equivalently φ is weakly conformal. In the work of Calabi [24], Eells-Wood [34], Ramanathan [59] and others, more complicated holomorphic differentials are constructed to establish isotropy properties for harmonic maps into spheres, projective spaces and Grassmannians. In this section we shall construct a large class of such holomorphic differentials for any symmetric space.

Our treatment is a Lie-theoretic translation of a construction of Simon Salamon in the case that $G = \mathbf{U}(n)$ which he obtained from the tournaments approach [21] to studying harmonic maps into Grassmannians.

We shall begin in a very general context. Let M be a Riemann surface, $\mathbf{g}$ a semisimple Lie algebra, $\underline{\mathbf{g}}$ the trivial bundle over M, $D = d - \mathrm{ad}\beta$ the connection in $\underline{\mathbf{g}}$ defined by a $\mathbf{g}$-valued 1-form β satisfying $d\beta = [\beta \wedge \beta]$. (Of course, we have in mind the pull-back of the canonical connection of some symmetric space.) Equip $\underline{\mathbf{g}}^{\mathbf{C}}$ with the Koszul-Malgrange holomorphic structure determined by D. Let $\mathbf{q}$ be a bundle of parabolic subalgebras of $\mathbf{g}^{\mathbf{C}}$ with nilradical bundle $\mathbf{n}$ and canonical section Ξ. We now have

Proposition 6.12. *$\mathbf{q}$ is a holomorphic sub-bundle if and only if $D_X \Xi$ is in $C^\infty(\mathbf{n})$ for all* (0,1) *vector fields X.*

Proof. We decompose $\underline{\mathbf{g}}^{\mathbf{C}}$ into a sum of eigen-bundles $\mathbf{g}_k$ of $\mathrm{ad}\Xi$ with eigenvalues $\sqrt{-1}k$ so that

$$\mathbf{q} = \sum_{k \geq 0} \mathbf{g}_k , \qquad \mathbf{n} = \sum_{k > 0} \mathbf{g}_k .$$

If $\eta \in C^\infty(\mathbf{g}_k)$ then

$$[\Xi, \eta] = \sqrt{-1}k\eta$$

and covariant differentiation by $X \in C^\infty(T^{0,1}M)$ gives

$$\sqrt{-1}kD_X\eta = [\Xi, D_X\eta] + [D_X\Xi, \eta]$$

from which it follows that $D_X\eta$ has values in $\mathbf{g}_k + [D_X\Xi, \mathbf{g}_k]$. Hence if $D_X\Xi \in C^\infty(\mathbf{n})$, $D_X C^\infty(\mathbf{q}) \subset C^\infty(\mathbf{q})$ and $\mathbf{q}$ is therefore holomorphic.

Conversely, if $\mathbf{q}$ is holomorphic, $D_X\Xi$ lies in $C^\infty(\mathbf{q})$ since Ξ does. However, Ξ has image in a single conjugacy class in $\mathbf{g}$ from which it follows that $d\Xi$ has values in $[\Xi, \underline{\mathbf{g}}^{\mathbf{C}}]$. Thus $D_X\Xi$ has no component in $\mathbf{g}_0$ and so lies in $C^\infty(\mathbf{n})$. □

Let $\mathbf{q}\subset\mathbf{g}^{\mathbf{C}}$ be a holomorphic bundle of parabolic subalgebras with canonical section $\Xi\in C^\infty(\mathbf{q})$. By (6.12), there is a unique (0,1)-form ε with values in $\mathbf{n}$ such that for $X\in T^{0,1}M$,

$$D_X\Xi = [\varepsilon(X), \Xi] \ . \tag{8}$$

Observe that this makes Ξ holomorphic for the Koszul-Malgrange structure determined by $D - \mathrm{ad}(\varepsilon+\bar{\varepsilon})$. In particular, each eigenbundle $\mathbf{g}_k$ is holomorphic with respect to this structure and so we have

Lemma 6.13. *If $\eta\in C^\infty(\mathbf{g}_k)$, $X\in T^{0,1}M$, then $D_X\eta-[\varepsilon(X), \eta]\in\mathbf{g}_k$.*

Now, for each k, set

$$\mathbf{g}^k = \sum_{j\geq k}\mathbf{g}_j \ .$$

Then $(\mathbf{g}^k)^\perp = \mathbf{g}^{1-k}$ and, for positive k, $\mathbf{g}^k$ is just the kth step in the central descending series for $\mathbf{n}$ so that each $\mathbf{g}^k$ is holomorphic. In particular, $\mathbf{g}^k/\mathbf{g}^{k+1}$ is also holomorphic. Of course, this last is C^∞-isomorphic to $\mathbf{g}_k$.

Now let z be a local holomorphic co-ordinate on M, set $\delta=\beta(\frac{\partial}{\partial z})$, $\varepsilon=\varepsilon(\frac{\partial}{\partial \bar{z}})$ and let ∂, $\bar{\partial}$ denote covariant differentiation by $\frac{\partial}{\partial z}$ and $\frac{\partial}{\partial \bar{z}}$ respectively.

Lemma 6.14. $\partial\varepsilon - \bar{\partial}\bar{\varepsilon} + [\varepsilon, \bar{\varepsilon}] + [\delta, \bar{\delta}]$ *takes values in* $\mathbf{g}_0$.

Proof. Arguing as in (1.4), we see that D has curvature $-\frac{1}{2}\mathrm{ad}[\beta\wedge\beta]$ whence, using (8), we have

$$\begin{aligned} -[[\delta, \bar{\delta}], \Xi] &= [\partial, \bar{\partial}]\Xi = \partial[\varepsilon, \Xi] - \bar{\partial}[\bar{\varepsilon}, \Xi] \\ &= [\partial\varepsilon, \Xi] + [\varepsilon, [\bar{\varepsilon}, \Xi]] - [\bar{\partial}\bar{\varepsilon}, \Xi] - [\bar{\varepsilon}, [\varepsilon, \Xi]] \\ &= [\partial\varepsilon - \bar{\partial}\bar{\varepsilon} + [\varepsilon, \bar{\varepsilon}], \Xi] \end{aligned}$$

and the result follows. □

Let ε_j denote the component of ε in $\mathbf{g}_j$ and let p be the maximum value of j for which ε_j is non-zero. Thus

$$\varepsilon = \varepsilon_1 + \ldots + \varepsilon_p \ .$$

Proposition 6.15. $\bar{\partial}\bar{\varepsilon}_p - [\delta, \bar{\delta}]$ *takes values in* $\mathbf{g}^{1-p}$.

Proof. By (6.13), if η is a section of $\mathbf{g}_j$, then $\bar{\partial}\eta-[\varepsilon, \eta]$ lies in $\mathbf{g}_j$ also, whence $\bar{\partial}\eta$ takes values in $\mathbf{g}_j+\ldots+\mathbf{g}_{j+p}$. Conjugating now shows that $\partial\eta$ lies in $\mathbf{g}_j+\ldots+\mathbf{g}_{j-p}$. Applying this to the components of ε we get

$$\partial\varepsilon\in\mathbf{g}_p+\ldots+\mathbf{g}_{1-p}\subset\mathbf{g}^{1-p} \ ,$$

while, for $j<p$,

$$\overline{\partial\varepsilon_j}\in\mathbf{g}_{-j}+\ldots+\mathbf{g}_{p-j}\subset\mathbf{g}^{1-p}\ .$$

Lastly, ε has no component in $\mathbf{g}_0$ so that $[\varepsilon,\bar{\varepsilon}]$ lies in $\mathbf{g}^{1-p}$ while $\mathbf{g}^0\subset\mathbf{g}^{1-p}$. The result now follows from (6.14). □

Corollary 6.16. *If $[\delta,\bar{\delta}]$ has values in $\mathbf{g}^{1-p}$, then $\overline{\varepsilon_p}$ projects to a (local) holomorphic section of $\mathbf{g}^{-p}/\mathbf{g}^{1-p}$.*

Globally, of course, this means that $\overline{\varepsilon_p}$ determines a holomorphic 1-form with values in $\mathbf{g}^{-p}/\mathbf{g}^{1-p}$.

To apply (6.16), we must have $[\delta,\bar{\delta}]$ lying in $\mathbf{g}^{1-p}$. This will certainly be the case if δ has values in a single eigenbundle for then $[\delta,\bar{\delta}]$ lies in $\mathbf{g}_0$ which is contained in $\mathbf{g}^{1-p}$ for any positive p. We therefore have

Corollary 6.17. *If δ lies in a single eigen-bundle of the canonical section then $\overline{\varepsilon_p}$ projects to a local holomorphic section of $\mathbf{g}^{-p}/\mathbf{g}^{1-p}$.*

F. Applications

We now apply the results of the previous section to the bundle of parabolic subalgebras associated to a harmonic 2-sphere in a symmetric space by (6.10). In this case, we can show that there are no non-zero holomorphic differentials with values in $\mathbf{g}^{-p}/\mathbf{g}^{1-p}$ for positive p.

Lemma 6.18. *Let $\varphi:S^2\to N$ be a harmonic map into a symmetric space of compact type and let $\mathbf{Q}$ be the bundle of τ-maximal parabolic subalgebras constructed in (6.10). Then there are no non-zero holomorphic 1-forms with values in $\mathbf{g}^p/\mathbf{g}^{1-p}$ for positive p.*

Proof. Recall that if $\{E_i\}$ is the HN filtration of $\underline{\mathbf{g}}^{\mathbf{C}}$ then we have

$$E_1\subset\mathbf{Q}=\mathbf{g}^0$$

whence $\mathbf{g}^{\mathbf{C}}/\mathbf{g}^0$ consists of line bundles of non-positive degree. Thus, for $p>0$, $\sup\mathbf{g}^{-p}/\mathbf{g}^{1-p}\le 0$ while $\inf T^{1,0}M=2$ so that, by (2.7), there are no non-zero holomorphic morphisms $T^{1,0}M\to\mathbf{g}^{-p}/\mathbf{g}^{1-p}$. □

We can now prove our main vanishing theorem. In the situation of (6.10), δ is just $\varphi_*(\frac{\partial}{\partial z})$ and, by (6.17), if δ lies in a single eigenbundle of the canonical section of $\mathbf{Q}$, $\overline{\varepsilon_p}$ is holomorphic and hence vanishes by (6.18). Thus ε and $\bar{\varepsilon}$ vanish identically and we conclude that the canonical section is D-parallel. To summarise:

Theorem 6.19. *Let $\varphi:S^2\to N$ be a harmonic map into a symmetric space of compact type and let $\mathbf{Q}$ be as in (6.10) with canonical section Ξ. If $\varphi_*(T^{1,0}M)$ lies in a single eigenbundle of Ξ then Ξ is parallel with respect to the pull-back of the canonical connection of N.*

There is one circumstance where the hypothesis on $d\varphi$ is always satisfied: that is when $\mathbf{Q}$ determines a J_2-holomorphic map into a flag manifold of height no greater than two canonically fibred over N. For then,

$$\mathbf{Q}\cap\varphi^{-1}[\mathbf{p}^{\mathbf{C}}]=\mathbf{g}_1$$

so that $d\varphi(T^{1,0}M)\subset \mathfrak{g}_1$ and (6.19) applies. Thus, in this case, Ξ and hence $\underline{\psi}$ are parallel and so, by (2.3), ψ is horizontal and J_1-holomorphic. In particular, this gives us

Theorem 6.20. *Let N be a compact inner symmetric space for which all canonically fibred flag manifolds have height no greater than two. Let* $\varphi: S^2 \to N$ *be a harmonic map. Then the lift* ψ *of (6.11) is* J_1*-holomorphic and horizontal.*

We recall from (4.26) and (5.11) that S^{2n} and $\mathbf{C}P^n$ only admit such flag manifolds and so (6.20) applies. Thus any harmonic map of a 2-sphere into S^{2n} and $\mathbf{C}P^n$ is covered by a horizontal holomorphic map into a canonically fibred flag manifold. This provides a uniform proof of results of Calabi [24] (S^{2n}), Bryant [10] (S^4 considered as the $\mathbf{Sp}(2)$-symmetric space $\mathbf{H}P^1$) and Eells-Wood [34] ($\mathbf{C}P^n$).

Remarks. (i) In the situation of (6.11), it is straight-forward to check that $\varepsilon+\bar{\varepsilon}$ is just the vertical component of $d\psi$.

(ii) A case by case calculation shows that even-dimensional spheres and complex projective spaces are the only compact inner irreducible symmetric spaces for which all canonically fibred flag manifolds have height not exceeding two.

Chapter 7. Stable Harmonic 2-spheres

In this chapter we shall give the proofs of the results announced in [20] as an application of the techniques we have been developing. This is joint work with Simon Salamon and we are grateful to him for allowing us to present this material here.

The main results are a complete determination of the stable harmonic maps of a 2-sphere into an irreducible Riemannian symmetric space.

A. Second variation formulae

Let $\varphi:(M,g)\to(N,h)$ be a smooth map between Riemannian manifolds. In chapter 2 we described what it means for φ to be a harmonic map: namely it is a critical point for the energy functional $E(\varphi)$. We now want to examine the stability of such maps. More precisely we shall say that φ is stable if the second variation $\nabla^2 E$ is a non-negative bilinear form on the space of variations $C^\infty(\varphi^{-1}TN)$. This is a weak notion of stability, but it will prove sufficient for our purposes.

The second variation of the energy was determined by Smith [68] who obtained the following formula (henceforth, the summation convention is in force):

$$\nabla^2_\varphi E\,(u,v) = \int_M \{h(\nabla^\varphi{}_{X_i}u, \nabla^\varphi{}_{X_i}v) + h(R^N(\varphi_*X_i,u)\varphi_*X_i,v)\}\, dvol_M\ . \tag{1}$$

where $u,v\in C^\infty(\varphi^{-1}TN)$, X_i, $i=1,\ldots,\dim M$ is a local orthonormal frame field on M and R^N denotes the Riemann curvature tensor of N.

We shall eventually be interested in the case where M is a Riemann surface, but initially we only assume (M,g,J) is an almost Hermitian manifold and rewrite (1) in terms of a local unitary frame field Z_j for $T^{(1,0)}M$. Setting

$$Z_j = 1/\sqrt{2}(X_j - \sqrt{-1}JX_j)\ ,$$

and writing ∇ for ∇^φ formula (1) becomes

$$\begin{aligned}\nabla^2_\varphi E(u,v) = 2\int_M \{&h(\nabla_{Z_j}u,\nabla_{\bar Z_j}v) + h(\nabla_{\bar Z_j}u,\nabla_{Z_j}v) + \\ &h(R^N(\varphi_*Z_j,u)\varphi_*\bar Z_j,v) + h(R^N(\varphi_*\bar Z_j,u)\varphi_*Z_j,v)\}\, dvol_M\ .\end{aligned} \tag{2}$$

Now, for $Z\in T^{1,0}M$,

$$\begin{aligned} h(\nabla_Z u,\nabla_{\bar Z}v) &= \bar Z h(\nabla_Z u,v) - h(\nabla_{\bar Z}\nabla_Z u,v) \\ &= \bar Z h(\nabla_Z u,v) - h(\nabla_Z\nabla_{\bar Z}u,v) - h(\nabla_{[\bar Z,Z]}u,v) - h(R^N(\varphi_*\bar Z,\varphi_*Z)u,v) \\ &= \{\bar Z h(\nabla_Z u,v) - Zh(\nabla_{\bar Z}u,v) - h(\nabla_{[\bar Z,Z]}u,v)\} + \\ &\quad h(\nabla_{\bar Z}u,\nabla_Z v) - h(R_N(\varphi_*\bar Z,\varphi_*Z)u,v)\ . \end{aligned} \tag{3}$$

Let us define a 1-form α on M by

$$\alpha(X) = h(\nabla_X u,v)\ .$$

Then (3) reads

$$h(\nabla_Z u,\nabla_{\bar{Z}} v) = h(\nabla_{\bar{Z}} u,\nabla_Z v) + d\alpha(\bar{Z},Z) - h(R_N(\varphi_*\bar{Z},\varphi_* Z)u,v)\ .$$

Substituting this into (2) and using the first Bianchi identity gives

$$\nabla^2_\varphi E(u,v) = 2\int_M d\alpha(\bar{Z}_j,Z_j) + 2h(\nabla_{\bar{Z}_j} u,\nabla_{Z_j} v) + 2h(R^N(\varphi_* Z_j,u)\varphi_*\bar{Z}_j,v)\ dvol_M\ .$$

Moreover, if ω^M is the Kähler form of M,

$$d\alpha(\bar{Z}_j,Z_j) = \sqrt{-1}h(d\alpha,\omega^M)$$

so that integration by parts gives

$$\int_M d\alpha(\bar{Z}_j,Z_j)\ dvol_M = \sqrt{-1}\int_M h(\alpha,d^*\omega^M)\ dvol_M$$

which last vanishes when ω^M is co-closed. We have thus proved

Theorem 7.1. (c.f. Micallef-Moore [52]) *Let (M,J,g) be an almost Hermitian manifold with co-closed Kähler form and $\varphi{:}M \to (N,h)$ a harmonic map. Then the second variation of the energy at φ is given by*

$$\nabla^2_\varphi E(u,v) = 4\int_M h(\nabla_{\bar{Z}_j} u,\nabla_{Z_j} v) + h(R^N(\varphi_* Z_j,u)\varphi_*\bar{Z}_j,v)\ dvol_M$$

where $\{Z_j\}$ is a local unitary frame for $T^{1,0}M$.

We now specialise to the case where M is a Riemann surface. If z is a holomorphic co-ordinate on M then the metric g has the form $\lambda^2|dz|^2$ for some positive function λ so that a unitary frame for $T^{1,0}M$ is given by $\lambda^{-1}\dfrac{\partial}{\partial z}$ while $dvol_M = \lambda^2 d^2z$. Thus, in this case, (7.1) reads

$$\nabla^2_\varphi E(u,v) = 4\int_M h(\nabla_{\frac{\partial}{\partial\bar{z}}} u,\nabla_{\frac{\partial}{\partial z}} v) + h(R^N(\delta,u)\bar{\delta},v)\ d^2z \tag{4}$$

where, as usual, $\delta = \varphi_*(\dfrac{\partial}{\partial z})$.

We extend (4) by complex bilinearity to all sections of $\varphi^{-1}TN^{\mathbf{C}}$ and then φ is stable if and only if

$$\nabla^2_\varphi E(u,\bar{u}) \geq 0$$

for all $u \in C^\infty(\varphi^{-1}TN^{\mathbf{C}})$. Equipping $\varphi^{-1}TN^{\mathbf{C}}$ with the Koszul-Malgrange structure from ∇, a holomorphic section of $\varphi^{-1}TN^{\mathbf{C}}$ has

$$\nabla_{\frac{\partial}{\partial\bar{z}}} u = 0$$

so that the first term in the integrand of (4) vanishes and we conclude

Theorem 7.2. *Let $\varphi{:}M \to (N,h)$ be a stable harmonic map of a Riemann surface and u a holomorphic section of $\varphi^{-1}TN^{\mathbf{C}}$ relative to the Koszul-Malgrange holomorphic structure. Then*

$$\int_M h(R^N(\delta,u)\bar{\delta},\bar{u})\ d^2z \geq 0\ .$$

Now let us take N to be a Riemannian symmetric G-space of compact type with metric induced by an invariant bilinear form $(.,.)$ on $\mathbf{g}$. We identify $\varphi^{-1}TN^{\mathbf{C}}$ with $\varphi^{-1}[\mathbf{p}^{\mathbf{C}}] \subset \underline{\mathbf{g}}^{\mathbf{C}}$ via the Maurer-Cartan form and then, from (1.4),

$$h(R^N(\delta,u)\bar{\delta},\bar{u}) = -([[\delta,u],\bar{\delta}],\bar{u}) = -|[\delta,u]|^2$$

and we have

Corollary 7.3. *If $\varphi:M \to N$ is a stable harmonic map of a Riemann surface into a Riemannian symmetric space of compact type and δ is viewed as a (local) section of $\varphi^{-1}[\mathbf{p}^{\mathbf{C}}]$ then $[\delta, u] = 0$ for all holomorphic sections of $\varphi^{-1}[\mathbf{p}^{\mathbf{C}}]$.*

When M is a Riemann sphere we can draw a much stronger conclusion because the Birkhoff-Grothendieck theorem (6.7) allows us to pass from statements about holomorphic sections to statements about holomorphic bundles. Indeed, denote $\varphi^{-1}[\mathbf{p}^{\mathbf{C}}]$ as a holomorphic bundle by T and let T_0 be the corresponding subbundle in the Harder-Narasimhan filtration. Then T_0 is a sum of line bundles of non-negative degree and so is spanned fibre-wise by its holomorphic sections. Thus we obtain

Corollary 7.4. *If $\varphi:S^2 \to N$ is a stable harmonic map into a Riemannian symmetric space of compact type then δ commutes with the fibres of T_0.*

B. Stability of harmonic 2-spheres

It is now convenient to assume that N is simply connected and irreducible. That this involves no loss of generality can be seen by the following argument. Firstly, as we are considering maps of S^2, we may take N to be simply connected since it is easy to check that a map of S^2 into N is stable harmonic if and only if any lift of that map into the universal cover of N is stable harmonic. Thus we may suppose that N is a Riemannian product of simply connected irreducible symmetric spaces and we may treat each factor separately.

Let us now apply the results of chapter 6 to further analyse the condition on φ given by (7.4). We denote by E the trivial bundle $\underline{\mathbf{g}}^{\mathbf{C}}$ equipped with its Koszul-Malgrange holomorphic structure and let

$$\ldots \subset E_1 \subset E_0 \subset E_{-1} \subset \ldots$$

be its Harder-Narasimhan filtration. In (6.10) we constructed a holomorphic bundle of τ-maximal parabolic subalgebras $\mathbf{Q} \subset E$ with nilradical bundle $\mathbf{N}$ such that

$$\varphi_*(T^{1,0}S^2) \subset \mathbf{N} \cap \varphi^{-1}[\mathbf{p}^{\mathbf{C}}] \subset \mathbf{Q} \cap \varphi^{-1}[\mathbf{p}^{\mathbf{C}}] \subset E_0 \cap \varphi^{-1}[\mathbf{p}^{\mathbf{C}}] = T_0 \ . \tag{5}$$

Thus δ has values in $\mathbf{N}$ and commutes with $\mathbf{Q} \cap \varphi^{-1}[\mathbf{p}^{\mathbf{C}}]$ by (7.4).

As an immediate application of this, we have

Theorem 7.5. *Let G be a compact semi-simple Lie group and $\varphi:S^2 \to G$ a stable harmonic map. Then φ is constant.*

Proof. We are of course viewing G as a $G \times G$-symmetric space with τ the standard involution on $\mathbf{g} \times \mathbf{g}$. From the above arguments, it suffices to prove that if $\mathbf{q}$ is a τ-maximal parabolic subalgebra of $(\mathbf{g} \times \mathbf{g})^{\mathbf{C}}$ then no element of $\mathbf{n}_{\mathbf{p}}$ centralises $\mathbf{q}_{\mathbf{p}}$. However, $\mathbf{q}$ contains a fundamental CSA $\mathbf{t}^{\mathbf{C}}$ for which, in this case, $\mathbf{t}_{\mathbf{p}}^{\mathbf{C}}$ is maximal abelian in $\mathbf{p}^{\mathbf{C}}$ so that any element of $\mathbf{p}^{\mathbf{C}}$ that centralises $\mathbf{q}_{\mathbf{p}}$ is contained in $\mathbf{t}_{\mathbf{p}}^{\mathbf{C}}$ and hence is semi-simple. This concludes the proof. □

Remarks. (i) The above argument establishes the triviality of stable harmonic 2-spheres in any *split-rank* symmetric space i.e. any compact symmetric space G/K satisfying

$$\operatorname{rank} G = \operatorname{rank} K + \operatorname{rank} G/K \ .$$

(ii) Theorem (7.5) can also be proved by quite different arguments. For example, Uhlenbeck has constructed an explicit negative variation for a non-constant 2-sphere in any compact Lie group.

Having disposed of the type II symmetric spaces, we turn our attention to those of type I and so take $\mathbf{g}$ to be simple and of compact type. We begin by drawing another, more fundamental, consequence from (5). Indeed, by the definition of τ-maximality, we know that $\mathbf{N}$ is generated by the $[\mathbf{p}^{\mathbf{C}}]$-parts of $\mathbf{N}$ and $\mathbf{Q}$ so that if δ centralises these, it must take values in the centre of $\mathbf{N}$ and we have

Theorem 7.6. *Let $\varphi: S^2 \to N$ be a stable harmonic map into a symmetric G-space of compact type and $\mathbf{Q} \subset \mathbf{g}^{\mathbf{C}}$ the bundle of parabolic subalgebras given by (6.10). Then $\varphi_*(T^{1,0}S^2)$ lies in the centre of the nilradical bundle of* $\mathbf{Q}$.

Let $\mathbf{Z}$ be the centre of $\mathbf{N}$. Since $\mathbf{g}$ is simple, (4.30) tells us that $\mathbf{Z}$ must lie wholly in $\varphi^{-1}[\mathbf{p}^{\mathbf{C}}]$ or $\varphi^{-1}[\mathbf{k}^{\mathbf{C}}]$. Clearly, if the latter holds then δ must vanish and φ be constant. Now each fibre of $\mathbf{Z}$ contains the highest root space of $\mathbf{g}^{\mathbf{C}}$ with respect to a suitable Weyl chamber of a fundamental CSA and so, in particular, contains a root space corresponding to a long root. Thus, in the language of chapter 3, if $\mathbf{Z} \subset \varphi^{-1}[\mathbf{p}^{\mathbf{C}}]$, there is a long $I_{\mathbf{p}}$ root. Now (3.27) tells us that $\pi_2(N) = 0$ if and only if any $I_{\mathbf{p}}$ roots are short so that we have, after taking (7.5) into account,

Theorem 7.7. *Let $\varphi: S^2 \to N$ be a stable harmonic map into a simply connected Riemannian symmetric space with $\pi_2(N) = 0$. Then φ is constant.*

Remark. A case by case check reveals that the list of type I symmetric spaces with vanishing π_2 coincides with the list of type I symmetric spaces with unstable identity map. Howard-Wei, Ohnita and Plushnikov [45, 56, 57] have shown that an irreducible symmetric space with unstable identity map is neither the range or domain of a non-constant stable harmonic map so that (7.7) follows from their work for N of type I. However, our proof is a priori. It is also interesting to note that for type II spaces, (7.7) does not so follow as there are simple Lie groups with stable identity map.

We now consider the case that N is Hermitian symmetric. Siu and Zhong [66, 87] have shown that any stable harmonic 2-sphere in a compact irreducible Hermitian symmetric space is $\pm$holomorphic. Their proof involves a case by case analysis of the curvature tensor of the Hermitian symmetric spaces. We now present an a priori proof of their result.

So let $\mathbf{H} = \mathbf{Q} \cap \underline{\mathbf{g}}$ and suppose that $N = G/K$ is irreducible Hermitian symmetric. Then we have a parallel splitting $T = T' \oplus T''$ induced by the type decomposition on N which is $\operatorname{ad} \mathbf{H}$ invariant since $\mathbf{H} \subset \varphi^{-1}[\mathbf{k}^{\mathbf{C}}]$ for inner symmetric spaces. This induces an $\operatorname{ad} \mathbf{H}$ invariant splitting of $\mathbf{Z}$:

$$\mathbf{Z} = \mathbf{Z} \cap T' \oplus \mathbf{Z} \cap T'' .$$

Since $\mathbf{g}$ is simple, by (4.3) $\mathbf{H}$ acts fibrewise irreducibly on $\mathbf{Z}$ and so we conclude that one of $\mathbf{Z} \cap T'$, $\mathbf{Z} \cap T''$ must vanish. Thus φ is holomorphic or anti-holomorphic and we have

Theorem 7.8. *If $\varphi: S^2 \to N$ is a stable harmonic map into a compact irreducible Hermitian symmetric space then φ is holomorphic or anti-holomorphic.*

An alternative formulation of this result is that a stable harmonic 2-sphere in any compact Hermitian symmetric space is holomorphic for at least one invariant complex structure.

C. Holomorphic factorisation

It remains to treat those symmetric spaces with $\pi_2(N) = \mathbf{Z}_2$. This requires further analysis of the subbundle $\mathbf{Z}$.

Let Ξ be the canonical section of $\mathbf{Q}$. By (4.3) (we are still assuming that $\mathbf{g}$ is simple), $\mathbf{Z}$ and $\overline{\mathbf{Z}}$ are both eigenbundles of Ξ with eigenvalues of opposite signs. In particular, $\varphi_*(T^{1,0}S^2)$ lies in a single eigenbundle and we have from (6.19) that Ξ is parallel. We now define $\mathbf{A} \subset T$ by

$$\mathbf{A} = (\mathbf{Z} + \overline{\mathbf{Z}}) \cap \varphi^{-1}[\mathbf{p}] .$$

Since all fibres of $\mathbf{Q}$ and hence of $\mathbf{A}$ are $\mathrm{Ad}(G)$-conjugate, $\mathbf{A}$ is a subbundle of T which in view of (4.30) is either zero or the real part of $\mathbf{Z}$.

Lemma 7.9. *Each fibre of* $\mathbf{A}$ *is a Lie triple system i.e*

$$[[\mathbf{A}, \mathbf{A}], \mathbf{A}] \subset \mathbf{A} .$$

Proof. Since $\mathbf{Z}$ is abelian,

$$[\mathbf{A}, \mathbf{A}] \subset [\mathbf{Z}, \overline{\mathbf{Z}}] \cap \varphi^{-1}[\mathbf{k}] .$$

Moreover, $\mathbf{Z}$ and $\overline{\mathbf{Z}}$ are eigenbundles of Ξ with eigenvalues that only differ in sign so that $[\mathbf{Z}, \overline{\mathbf{Z}}]$ lies in the zero eigenbundle $\mathbf{H}$. From this we see that

$$[[\mathbf{Z}, \overline{\mathbf{Z}}], \mathbf{Z}] \subset \mathbf{Z}$$

whence

$$[[\mathbf{Z}, \overline{\mathbf{Z}}] \cap \varphi^{-1}[\mathbf{k}], \mathbf{A}] \subset \mathbf{A}$$

and the lemma follows. □

In view of the well-known correspondence between Lie triple systems and totally geodesic subspaces (c.f. [42]), it follows that each $\mathbf{A}_x$ is the tangent space at $\varphi(x)$ of a totally geodesic symmetric subspace of N passing through $\varphi(x)$. Denote this subspace by N_x. We shall shortly see that φ has image in such a subspace.

Lemma 7.10. *For any vector field* X *on* S^2 *and section* η *of* $\mathbf{A}$, $\nabla_X \eta$ *is a section of* $\mathbf{A}$.

Proof. We have seen that Ξ is ∇-parallel so that each eigenbundle of Ξ is ∇-stable. In particular, $\mathbf{Z}$ is ∇-stable and since $\varphi^{-1}[\mathbf{p}]$ is also, the conclusion follows. □

Corollary 7.11. *Fix* $x_0 \in S^2$, *then* φ *has image contained in* N_{x_0}.

Proof. $T_{\varphi(x_0)}N_{x_0}$ is just $\mathbf{A}_{x_0}$ so that $d\varphi_{x_0}$ has image in $T_{\varphi(x_0)}N_{x_0}$. Moreover, by (7.10), all higher covariant derivatives of $d\varphi$ at x_0 also have values in $T_{\varphi(x_0)}N_{x_0}$ so that the image of φ osculates N_{x_0} to infinite order at x_0. We may now apply a Unique Continuation Theorem of Sampson [63] (or appeal to the real analyticity of everything in sight) to draw the conclusion. □

In fact we can do better than this. We shall now show that each N_x is an immersed Hermitian symmetric space through which φ factors holomorphically. For this, set $\mathbf{H}_{\mathbf{k}} = \mathbf{Q} \cap \varphi^{-1}[\mathbf{k}]$.

Proposition 7.12. $\mathbf{H}_{\mathbf{k}} + \mathbf{A}$ *is a constant subbundle of Lie algebras of the trivial bundle* $\underline{\mathbf{g}}$.

Proof. The arguments of (7.9) easily show that $\mathbf{H_k+A}$ is a bundle of Lie algebras. As for the constancy, recall that

$$\nabla = d - \mathrm{ad}\varphi^*\beta$$

where β is the Maurer-Cartan form of N. By (7.10) we know that $\mathbf{H_k+A}$ is ∇-stable while, since $\varphi^*\beta$ has values in $\mathbf{A}$, we have that $\mathbf{H_k+A}$ is $\mathrm{ad}\,\varphi^*\beta$-stable. We therefore conclude that $\mathbf{H_k+A}$ is stable under flat differentiation which proves the result. □

Set $\mathfrak{g}_0$ to be the Lie algebra given by any fibre of $\mathbf{H_k+A}$ and let G_0 be the corresponding analytic subgroup of G. For any $x \in S^2$, $\mathfrak{g}_0$ is stable under the involution of N at $\varphi(x)$ and so inherits a symmetric decomposition

$$\mathfrak{g}_0 = (\mathbf{H_k})_x + \mathbf{A}_x .$$

Moreover, the G_0-orbit of $\varphi(x)$ is just the totally geodesic submanifold N_x of N.

Remark. One can use (7.12) to give a proof of (7.11) without recourse to any unique continuation results. The idea is to show that φ is (locally) covered by maps into G for which the pull-back of the right Maurer-Cartan form has values in $\mathfrak{g}_0$. Then these lifts, up to right multiplication by a constant, have image in G_0 and thus φ has image in a G_0-orbit of N. However, for N non-inner, the construction of these lifts is somewhat lengthy. Now let us fix a base-point $x_0 \in S^2$ and let H_0 be the analytic subgroup of G with Lie algebra $(\mathbf{H_k})_{x_0}$.

Proposition 7.13. *G_0/H_0 is a Hermitian symmetric space and the map*

$$i_0 : gH_0 \mapsto g\cdot\varphi(x_0)$$

is an equivariant totally geodesic immersion into N with image N_{x_0}. Moreover, φ factors through G_0/H_0 holomorphically.

Proof. That G_0/H_0 is Hermitian symmetric follows immediately from the observation that the splitting

$$(\mathbf{A}_{x_0})^{\mathbf{C}} = \mathbf{Z}_{x_0} \cap \varphi^{-1}[\mathbf{p}^{\mathbf{C}}]_{x_0} \oplus \overline{\mathbf{Z}}_{x_0} \cap \varphi^{-1}[\mathbf{p}^{\mathbf{C}}]_{x_0}$$

is $\mathrm{Ad}(H_0)$-invariant. Moreover, $(\mathbf{H_k})_{x_0}$ is the Lie algebra of the stabiliser in G_0 of $\varphi(x_0)$ so that i_0 is well-defined and equivariant with respect to the inclusion of G_0 in G. It is now easy to see using (1.7) and (1.8) that i_0 is a totally geodesic immersion.

Thus G_0/H_0 is a covering space for N_{x_0} so that since S^2 is simply connected, φ factors through G_0/H_0. Finally, $\varphi_*(T^{1,0}S^2)$ is contained in $\mathbf{Z}$ and we conclude that φ factors holomorphically. □

We have therefore proved the following theorem:

Theorem 7.14. *Let $\varphi : S^2 \to N$ be a stable harmonic map into a Riemannian symmetric space of compact type. Then there is a totally geodesically immersed Hermitian symmetric space G_0/H_0 in N through which φ factors as a holomorphic map.*

Moreover, when N is irreducible of type I, the (1,0) vectors at $y \in G_0/H_0$ are given by the intersection of $[\mathbf{p}^{\mathbf{C}}]_{i_0(y)}$ with the centre of the nilradical of a $\tau_{i_0(y)}$-maximal parabolic subalgebra of $\mathfrak{g}^{\mathbf{C}}$.

We call the Hermitian symmetric spaces G_0/H_0 *stabilising subspaces* (see (7.16) below).

Remarks. *(i)* When $\pi_2(N)=\pi_1(N)=0$, we know that **A** vanishes and so the stabilising subspaces are just points whence (7.7) is a corollary of (7.14). Again, the argument in (7.8) shows that when N is irreducible Hermitian symmetric, the stabilising subspaces are $\pm$holomorphically immersed and we may also interpret (7.8) as a special case of (7.14).

(ii) Calculation of classical examples suggests that the stabilising subspaces are in fact embedded in N but we have not been able to find a proof of this.

(iii) Micallef-Moore [52] have proved the following result which is clearly related to (7.14):

If N is an even-dimensional Riemannian manifold whose sectional curvature is non-negative on totally isotropic 2-planes and $\varphi:S^2 \to N$ is stable and harmonic than there is a ∇-parallel almost complex structure J on $\varphi^{-1}TN$ such that

$$J\circ d\varphi = d\varphi\circ J_{S^2}$$

where J_{S^2} is the complex structure on S^2.

This result is proved by similar (but less complicated) methods. The force of (7.14) is that the J is in fact the pull-back of the complex structure on a stabilising subspace.

D. Stabilising subspaces

We now investigate the geometry of the stabilising subspaces G_0/H_0 of an irreducible symmetric G-space of type I. We begin by re-iterating the construction of such spaces. So let $x\in N$ with stabiliser K and involution τ. Then a stabilising subspace through x is built by the following procedure:

Let $\mathbf{q}$ be a τ-maximal parabolic subalgebra of $\mathbf{g}^{\mathbf{C}}$ and set $\mathbf{h}_0=\mathbf{q}\cap\mathbf{k}$. Let $\mathbf{z}$ be the centre of the nilradical of $\mathbf{q}$ (which we assume to be contained in $\mathbf{p}^{\mathbf{C}}$) and define $\mathbf{g}_0\subset\mathbf{g}^{\mathbf{C}}$ by

$$\mathbf{g}_0^{\mathbf{C}} = \mathbf{h}_0^{\mathbf{C}}+\mathbf{z}+\overline{\mathbf{z}}\ .$$

If G_0, H_0 are the corresponding analytic subgroups of G, then G_0/H_0 is the stabilising subspace and the homogeneous immersion $i_0:G_0/H_0 \to G/K$ is totally geodesic.

Proposition 7.15. *A stabilising subspace G_0/H_0 is a compact irreducible Hermitian symmetric space.*

Proof. We must realise G_0/H_0 as a Hermitian symmetric space G_1/H_1 where G_1 is compact and simple. For this, we define $\mathbf{g}_1\subset\mathbf{g}_0$ by

$$\mathbf{g}_1^{\mathbf{C}} = \mathbf{z}+\overline{\mathbf{z}}+[\mathbf{z},\overline{\mathbf{z}}]$$

and put

$$\mathbf{h}_1 = [\mathbf{z},\overline{\mathbf{z}}]\cap\mathbf{g} \qquad \mathbf{p}_1 = (\mathbf{z}+\overline{\mathbf{z}})\cap\mathbf{g}\ .$$

Then $\mathbf{g}_1$ is a Lie algebra with symmetric decomposition

$$\mathbf{g}_1 = \mathbf{h}_1\oplus\mathbf{p}_1$$

since $\mathbf{h}_1\subset\mathbf{k}$ and $\mathbf{p}_1\subset\mathbf{p}$. Now an element of $\mathbf{h}_0$ is orthogonal to $\mathbf{h}_1$ if and only if it centralises $\mathbf{p}_1$ so that $\mathbf{h}_1$ acts irreducibly on $\mathbf{z}$ and $\overline{\mathbf{z}}$ since $\mathbf{h}_0$ does by (4.30). In particular,

$$[\mathbf{h}_1,\mathbf{z}] = \mathbf{z}$$

so that $\mathbf{g}_1^{\mathbf{C}}$ is equal to its derived algebra and so is semi-simple. Indeed, using that fact that $\mathbf{h}_1$ acts irreducibly together with the identity

$$[\mathbf{p}_1, \mathbf{p}_1] = \mathbf{k}_1 \ ,$$

we may easily deduce that $\mathbf{g}_1$ is simple. Moreover $\mathbf{g}_1$ is a subalgebra of the compact algebra $\mathbf{g}$ from which we conclude (c.f. [74] pp. 345–6) that $\mathbf{g}_1$ is of compact type and the corresponding analytic subgroup G_1 is compact.

Consider now the G_1-orbit of eH_0 in G_0/H_0. This orbit is compact since G_1 is and open since $\mathbf{z}+\bar{\mathbf{z}}\subset\mathbf{g}_1^{\mathbf{C}}$ so that we conclude from the connectedness of G_0/H_0 that G_1 acts transitively on G_0/H_0. Thus $G_0/H_0 = G_1/H_1$ where $H_1 = H_0 \cap G_1$ and is therefore an irreducible Hermitian symmetric space. □

In fact we may identify G_1/H_1 by inspection of the extended Dynkin diagram of $\mathbf{g}$. For this we pick an ordering of the roots for which $\mathbf{q} = \mathbf{q}_I$ and let θ be the highest root for this ordering so that $\mathbf{g}^{\theta}\subset\mathbf{z}$. Then $\{\alpha_i\}_{i\notin I}\cup\{-\theta\}$ is a set of simple roots for the semi-simple part of $\mathbf{g}_0$. Now $\mathbf{g}_1$ is a simple ideal in the semi-simple part of $\mathbf{g}_0$ and contains $\mathbf{g}^{\theta}$ so the Dynkin diagram of $\mathbf{g}_1$ is the connected component in the extended Dynkin diagram of $\mathbf{g}_0$ which contains $-\theta$. In summary then, the Dynkin diagram of $\mathbf{g}_1$ is obtained by taking the extended Dynkin diagram of $\mathbf{g}$, striking out the roots in I and taking the connected component of what remains that contains $-\theta$. The results of chapter 3 may now be applied to find $\mathbf{h}_1$: since $-\theta$ is the only $I_{\mathbf{p}_1}$ simple root, we find from (3.17) that $\mathbf{h}_1$ has 1-dimensional centre and the Dynkin diagram of the semi-simple part of $\mathbf{h}_1$ is obtained from that of $\mathbf{g}_1$ by striking out $-\theta$.

Applying this analysis to a case by case study of extended Dynkin diagrams together with a case by case check of which τ-maximal parabolics have $\mathbf{z}\subset\mathbf{p}^{\mathbf{C}}$ leads one to conclude that if $\pi_2(N) = \mathbf{Z}_2$, then all the stabilising subspaces are complex projective spaces. For the exceptional such N, the maximal stabilising subspaces are listed in [20].

We have seen that any stable holomorphic 2-sphere in N factors holomorphically through a stabilising subspace. The most remarkable property of such subspaces is that a converse to this is true:

Theorem 7.16. *Let $i{:}G_1/H_1 \to N$ be a stabilising subspace of a compact irreducible symmetric space and M an almost Hermitian manifold with co-closed Kähler form. If $\varphi{:}M \to G_1/H_1$ is $\pm$holomorphic then $i\circ\varphi{:}M \to N$ is a stable harmonic map.*

Proof. By a result of Lichnerowicz [51] φ is harmonic whence $i\circ\varphi$ is since i is totally geodesic. It remains to prove the stability assertion. For this we begin by noting that $i^{-1}TN^{\mathbf{C}}$ has a G_1-invariant splitting

$$i^{-1}[\mathbf{p}^{\mathbf{C}}] = [\mathbf{n}_{\mathbf{p}}] + [\mathbf{a}_{\mathbf{p}}] + [\bar{\mathbf{n}}_{\mathbf{p}}] \qquad (6)$$

where, of course, $\mathbf{n}_{\mathbf{p}}$ is the $\mathbf{p}$-part of the nilradical of our τ-maximal parabolic subalgebra $\mathbf{q}$ and $\mathbf{a}_{\mathbf{p}}$ is the $\mathbf{p}$-part of the reductive part of $\mathbf{q}$. We also observe that the Maurer-Cartan form of G_1/H_1 identifies $T^{1,0}G_1/H_1$ with $[\mathbf{z}]\subset[\mathbf{n}_{\mathbf{p}}]$ while the pull-back by i of the canonical connection on N is just the canonical connection on G_1/H_1 since i is totally geodesic.

Suppose now that $\varphi{:}M \to G_1/H_1$ is a holomorphic map. Define a Hermitian form on $C^{\infty}((i\circ\varphi)^{-1}TN^{\mathbf{C}})$ by

$$\mathrm{Hess}(u,v) = \nabla^2_{i\circ\varphi}E(u,\bar{v}) \ .$$

We must show that Hess is positive semi-definite. If $Z_1,\dots,Z_m$ is a local unitary frame on M, we obtain the following formula for Hess from (7.1):

$$\mathrm{Hess}(u,v) = 4\int_M \langle \nabla_{\bar{Z}_j} u, \nabla_{\bar{Z}_j} v\rangle - \langle [\varphi_* Z_j, u],[\varphi_* Z_j, v]\rangle \, dvol_M \tag{7}$$

where $\langle .,. \rangle$ is the Hermitian inner product on $(i\circ\varphi)^{-1}[\mathbf{p}^{\mathbf{C}}]$. Since the decomposition (6) is H_1-invariant, its pull-back by φ is ∇-stable. Moreover, $\varphi_* Z_j$ has values in $\varphi^{-1}[\mathbf{z}]$ and so commutes with $\varphi^{-1}[\mathbf{a_p}+\mathbf{n_p}]$. From this it is clear that Hess diagonalises over the decomposition

$$C^\infty(\varphi^{-1}[\mathbf{n_p}]) + C^\infty(\varphi^{-1}[\mathbf{a_p}]) + C^\infty(\varphi^{-1}[\bar{\mathbf{n}}_\mathbf{p}]) \; .$$

Further, we see from (7) that for $u,v\in C^\infty(\varphi^{-1}[\mathbf{n_p}]+\varphi^{-1}[\mathbf{a_p}])$,

$$\mathrm{Hess}(u,v) = 4\int_M \langle \nabla_{\bar{Z}_j} u, \nabla_{\bar{Z}_j} v\rangle \, dvol_M$$

so that since Hess is real we may conclude that it is positive semi-definite on all of $C^\infty([\mathbf{p}^{\mathbf{C}}])$. □

Remarks. (i) In case that N is Hermitian symmetric, we note that the maximal stabilising subspace of N is just N itself.

(ii) Applying (7.16) to the identity map of G_1/H_1, we conclude that i is a stable harmonic map.

It remains to show that there is a sufficient supply of stabilising subspaces in the case when $\pi_2(N)=\mathbf{Z}_2$. For this we recall the root sphere construction of chapter 3, section E. Recall that any $I_\mathbf{p}$ root α gives rise to a totally geodesic map $\varphi_\alpha : S^2 \to N$. We now show that if α is a *long* $I_\mathbf{p}$ root then φ_α is stable. To do this it is sufficient to construct a τ-maximal parabolic subalgebra with $\mathbf{g}^\alpha$ contained in the centre of the nilradical, for then φ_α will factor holomorphically through the corresponding stabilising subspace and so be stable by (7.16).

So let α be a long $I_\mathbf{p}$ root. Then there is a Weyl chamber for which α is the highest root. Indeed, one can assume that this Weyl chamber is τ-stable (for almost every ξ in a Weyl chamber the $\mathbf{k}$-part of ξ also lies in a Weyl chamber since we are using a fundamental CSA). We then use the corresponding Borel subalgebra to construct a τ-maximal parabolic subalgebra $\mathbf{q}$ as in (4.29ii). As $\mathbf{g}^\alpha$ centralises the nilradical of the Borel subalgebra it certainly lies in $\mathbf{z}$ and we have shown

Proposition 7.17. *If $\alpha\in I_\mathbf{p}$ is long, the root sphere $\varphi_\alpha : S^2 \to N$ is stable.*

Recall from (3.29) that the long root spheres generate $\pi_2(N)$ while pre-composition by a holomorphic self-map of S^2 preserves harmonicity and, by (7.16), stability. Thus, as a corollary to (7.17), we finally conclude

Theorem 7.18. *If N is a simply connected irreducible Riemannian symmetric space, every homotopy class of maps $S^2 \to N$ has a stable harmonic representative.*

Remark. In case $\pi_2(N)=\mathbf{Z}_2$, observe that the above argument shows the existence of a *non-trivial* stable harmonic representative in the trivial homotopy class.

Chapter 8. Factorisation of harmonic spheres in Lie groups

There has been much recent interest in the construction and classification of harmonic 2-spheres in symmetric spaces (for a survey, see [17]). One approach to this problem was initiated by Uhlenbeck [72] who studied harmonic 2-spheres in the unitary group. In this chapter, we shall extend Uhlenbeck's results to other Lie groups using a method due to Valli [73] in the case of the unitary group.

A. Preliminaries

In previous chapters, we have considered harmonic maps into symmetric spaces. By (1.9), any such map gives rise to a harmonic map into a Lie group (the isometry group of the symmetric space) and it is to the study of such maps that we now turn.

We begin by returning to the construction of (1.9). Recall that if G is a Lie group and N a symmetric G-space, we may fix a base-point $x_0 \in N$ with involution τ_0 and then define a map $\Phi : N \to G$ by

$$\Phi(g.x_0) = g^{\tau_0} g^{-1} .$$

Φ is equivariant with respect to the homomorphism $\rho : G \to G \times G$, $g \mapsto (g^{\tau_0}, g)$, and is a totally geodesic immersion by (1.9). It is easy to see that changing the base-point changes Φ by left multiplication by a constant.

Lemma 8.1. *Let β^N be the Maurer-Cartan form of N and θ the left Maurer-Cartan form of G. Then*

$$\Phi^* \theta = -2\beta^N . \tag{1}$$

Proof. We begin by recalling the reductive homogeneous geometry of G as a symmetric $G \times G$-space. Let $\mathbf{m} = \{(\xi, -\xi) : \xi \in \mathbf{g}\}$. Then the symmetric decomposition of $\mathbf{g} \times \mathbf{g}$ at $g \in G$ is given by $[\Delta\mathbf{g}]_g \oplus [\mathbf{m}]_g$ and

$$[\mathbf{m}]_g = \{(\mathrm{Ad}(g)\xi, -\xi) : \xi \in \mathbf{g}\}$$

so that the projection P onto $[\mathbf{m}]$ along $[\Delta\mathbf{g}]$ is given by

$$P_g(\xi_1, \xi_2) = \tfrac{1}{2}(\xi_1 - \mathrm{Ad}(g)\,\xi_2, \xi_2 - \mathrm{Ad}(g^{-1})\,\xi_1) .$$

Now let β be the (symmetric) Maurer-Cartan form of G. Thus, if θ^R is the right Maurer-Cartan form of G we have

$$\beta = \tfrac{1}{2}(\theta^R, -\theta) .$$

By (1.8),

$$\Phi^* \beta = \Phi^{-1} P \circ \rho \circ \beta^N$$

and projecting onto the second factor gives

$$\Phi^* \theta = \mathrm{Ad}(\Phi^{-1}) \tau_0 \beta^N - \beta^N .$$

A straight-forward calculation shows that $\mathrm{Ad}(\Phi^{-1}(x))\tau_0$ is just the involution at $x \in N$ so that, since β_x^N has image in $[\mathbf{p}]_x$, we see that

$$\mathrm{Ad}(\Phi^{-1})\tau_0 \beta^N = -\beta^N$$

and the lemma is proved. □

In particular, if N and G inherit metrics from an invariant inner product on $\mathbf{g}$, we see that Φ is a homothety but not an isometry.

Let G be compact, connected and semi-simple. Let $\Gamma \subset \mathbf{g}$ denote the set of canonical elements of parabolic subalgebras of $\mathbf{g}^{\mathbf{C}}$. Then Γ is the disjoint union of a finite number of compact conjugacy classes each one of which is a flag manifold embedded in $\mathbf{g}$ by its canonical section.

Consider the map $\gamma : \Gamma \to G$ defined by $\xi \mapsto \exp \pi \xi$. Since γ is equivariant in the sense that

$$\gamma(\mathrm{Ad} g \xi) = g \gamma(\xi) g^{-1}$$

we see that the image of a component of Γ is a conjugacy class in G. Further, since $\mathrm{ad}\xi$ has eigenvalues in $\sqrt{-1}\mathbf{Z}$, $\gamma(\xi)^2$ lies in the centre of G so that conjugation by $\gamma(\xi)$ is an involution of G. Thus, if F is the conjugacy class of ξ, we see that $\gamma(F)$ is a totally geodesically immersed copy of the inner symmetric space $N(F)$. In fact, up to left translation by a constant, γ is just the composition of a canonical fibration of F with Φ. Now, if $\pi : F \to N$ is a canonical fibration and $P : \underline{\mathbf{g}} \to [\mathbf{p}]$ projection along $[\mathbf{k}]$, then, by (1.8), the Maurer-Cartan forms are related by

$$\pi^* \beta^N = P \beta^F$$

so that from (1) we get

$$\gamma^* \theta = -2P\beta^F = \mathrm{Ad}\gamma \beta^F - \beta^F \;. \tag{2}$$

Finally, suppose that $\Xi : M \to \Gamma$ is a map of a connected manifold M. Then Ξ has image in a single conjugacy class F and so may be viewed either as a map into the flag manifold F or, equivalently, as the (canonical section of) the corresponding bundle $\mathbf{q}^{\Xi} \subset \underline{\mathbf{g}}^{\mathbf{C}}$ of parabolic subalgebras.

Such a map induces a decomposition of $\underline{\mathbf{g}}^{\mathbf{C}}$ into a sum of eigen-bundles $\mathbf{g}_k$ of $\mathrm{ad}\Xi$ with eigenvalues $\sqrt{-1}k$ so that

$$\mathbf{q}^{\Xi} = \sum_{k \geq 0} \mathbf{g}_k$$

and so on. Further, the J_1 Kähler form on F pulls back to a 2-form ω^{Ξ} given by

$$\omega^{\Xi} = -\tfrac{1}{2}(\Xi, [\Xi^* \beta^F \wedge \Xi^* \beta^F])$$

where the inner product is the negative of the Killing form on $\mathbf{g}$.

B. Flag transforms

We now present a general procedure, essentially a Bäcklund transform, for producing new harmonic maps from a Riemann surface into a Lie group from old ones.

For simplicity of exposition, henceforth we make the following

Assumptions. G is a compact simple Lie group and M is a closed connected Riemann surface. Further, all symmetric G-spaces will be equipped with the metric induced by the negative of the Killing form while G itself will carry the Killing form metric pulled back by the left Maurer-Cartan form θ of G. Of course, any other invariant metric will be homothetic to these since G is simple.

Let $\varphi:M\to G$ be a map. We identify TG with $G\times\mathbf{g}$ via θ and then the pull-back of the Levi-Civita connection of G, ∇^{φ}, is given on $\underline{\mathbf{g}}$ by

$$\nabla^{\varphi} = d+\tfrac{1}{2}\mathrm{ad}\varphi^*\theta ,$$

(c.f. chapter 1). It is well-known (see e.g. [60]) that φ is harmonic if and only if $\varphi^*\theta$ is co-closed.

Definition. Let $\varphi:M\to G$. A *flag factor for* φ is a map $\Xi:M\to F\subset\Gamma$ such that

(*i*) $\varphi^*\theta^{1,0}$ has image in $\mathbf{g}_0+\mathbf{g}_1$;

(*ii*) $(\nabla^{\varphi}\Xi)^{0,1}$ has image in $\mathbf{g}_1$.

If Ξ is a flag factor for φ, we define a new map $\tilde{\varphi}:M\to G$ by

$$\tilde{\varphi} = \varphi\gamma(\Xi) = \varphi\exp\pi\Xi ,$$

where, of course, juxtaposition denotes pointwise multiplication in G. We call $\tilde{\varphi}$ the *flag transform of* φ *by* Ξ.

Condition (*ii*) has a geometrical interpretation. Indeed, if we equip $\underline{\mathbf{g}}^{\mathbf{C}}$ with the Koszul-Malgrange holomorphic structure given by ∇^{φ}, then, from (6.12), we have

Lemma 8.2. *If* $\Xi:M\to F$ *is a flag factor then* $\mathbf{q}^{\Xi}$ *is holomorphic. Further, if* F *has height one (equivalently,* F *is a Hermitian symmetric space), then (ii) is equivalent to the holomorphicity of* $\mathbf{q}^{\Xi}$.

Notation. For η a k-form with values in $\underline{\mathbf{g}}^{\mathbf{C}}$, let η_k denote the component of η with values in $\mathbf{g}_k$.

It will be useful in the sequel to phrase the flag factor equations for $\Xi:M\to F$ in terms of the Maurer-Cartan form β of F. Since the canonical section of F is invariant and therefore parallel for the canonical connection on F, we have

$$d\Xi = [\Xi^*\beta, \Xi] \tag{3}$$

so that

$$\nabla^{\varphi}\Xi = [\Xi^*\beta+\tfrac{1}{2}\varphi^*\theta, \Xi] . \tag{4}$$

From this, we see that (*ii*) is equivalent to

$$\Xi^*\beta_k^{1,0}+\tfrac{1}{2}\varphi^*\theta_k^{1,0} = 0 \qquad \text{for } k\neq 0,-1 , \tag{5}$$

which, together with (*i*), implies that $\Xi^*\beta$ has image in $\mathbf{g}_1+\mathbf{g}_{-1}$. Thus we have shown

Lemma 8.3. *If* $\Xi:M\to F$ *is a flag factor, then* Ξ *is super-horizontal.*

Of course, flag factors are not usually (anti-)holomorphic.

Energy and harmonicity of flag transforms

The importance of flag factors and the associated flag transforms comes from the following

Theorem 8.4. *Let* $\varphi:M\to G$ *be harmonic,* $\Xi:M\to F\subset\Gamma$ *a flag transform for* φ *and* $\tilde{\varphi}:M\to G$

the flag transform of φ by Ξ. Then $\widetilde{\varphi}$ is harmonic and

$$E(\varphi) - E(\widetilde{\varphi}) = -4\int_M \omega^{\Xi} .$$

Proof. Differentiating the product and using (2), we have

$$\begin{aligned}\widetilde{\varphi}^*\theta &= \mathrm{Ad}\gamma(\Xi)\varphi^*\theta + (\gamma(\Xi))^*\theta \\ &= \mathrm{Adexp}\pi\Xi\{\varphi^*\theta + \Xi^*\beta\} - \Xi^*\beta .\end{aligned}$$

On $\mathbf{g}_k$, $\mathrm{Adexp}\pi\Xi$ is just multiplication by $e^{\sqrt{-1}\pi k}$ so that by (*i*) and (8.3) we have

$$\widetilde{\varphi}^*\theta = \varphi^*\theta_0^{1,0} - \varphi^*\theta_1^{1,0} - 2\Xi^*\beta^{1,0} . \qquad (6)$$

We now use (3) and (5) to conclude

$$\widetilde{\varphi}^*\theta^{1,0} - \varphi^*\theta^{1,0} = 2\Xi^*\beta_1^{1,0} - 2\Xi^*\beta_{-1}^{1,0} = 2\sqrt{-1}\partial\Xi .$$

Thus

$$\widetilde{\varphi}^*\theta - \varphi^*\theta = 2\sqrt{-1}\partial\Xi - 2\sqrt{-1}\bar{\partial}\Xi = -2*d\Xi , \qquad (7)$$

where $*$ is the Hodge $*$-operator on M, whence

$$d^*\widetilde{\varphi}^*\theta = d^*\varphi^*\theta - 2*d^2\Xi = 0$$

and $\widetilde{\varphi}$ is harmonic.

As for the energy formula, first observe from (6) that

$$\widetilde{\varphi}^*\theta + \varphi^*\theta = 2\varphi^*\theta_0 - 2\Xi^*\beta$$

whence, using (7),

$$\begin{aligned}\{e(\widetilde{\varphi}) - e(\varphi)\}*1 &= \tfrac{1}{2}(\widetilde{\varphi}^*\theta + \varphi^*\theta, \widetilde{\varphi}^*\theta - \varphi^*\theta)*1 \\ &= 2(*d\Xi, \Xi^*\beta)*1 = 2(d\Xi \wedge \Xi^*\beta) \\ &= 4\omega^{\Xi} ,\end{aligned}$$

with the last inequality following from (3) and the invariance of the metric. Integrating over M now gives the result. □

Thus, by solving the first order flag transform equations (which are essentially constrained Cauchy-Riemann equations), we may build new harmonic maps from old. Moreover, this procedure is invertible.

Proposition 8.5. *Let $\varphi: M \to G$ be harmonic, $\Xi: M \to F \subset \Gamma$ a flag transform for φ and $\widetilde{\varphi}: M \to G$ the flag transform of φ by Ξ. Then $-\Xi: M \to \Gamma$ is a flag factor for $\widetilde{\varphi}$ so that φ is a flag transform of $\widetilde{\varphi}$.*

Proof. Clearly $-\Xi$ has image in Γ being the canonical section of the complex conjugate of $\mathbf{q}^{\Xi}$. As usual, we let $\mathbf{g}_k$ denote the eigen-bundles of $\mathrm{ad}\Xi$. Using (5) and (6), we have

$$\widetilde{\varphi}^*\theta^{1,0} = \varphi^*\theta_0^{1,0} - 2\Xi^*\beta_{-1}^{1,0} \qquad (8)$$

so that $\widetilde{\varphi}^*\theta^{1,0}$ has values in $\mathbf{g}_0 + \mathbf{g}_{-1}$ and $-\Xi$ satisfies condition (*i*).

Further,

$$(\nabla^{\widetilde{\varphi}}\Xi)^{1,0} = \partial\Xi + \tfrac{1}{2}[\widetilde{\varphi}^*\theta^{1,0}, \Xi]$$

which by (3) and (8) is just

$$[\Xi^*\beta^{1,0},\Xi] - [\Xi^*\beta^{1,0}_{-1},\Xi] = -\sqrt{-1}\Xi^*\beta^{1,0}_1 .$$

Thus $(\nabla^{\tilde{\varphi}}\Xi)^{0,1}$ takes values in $\mathbf{g}_{-1}$ so that $-\Xi$ satisfies condition (*ii*) and so is a flag transform for $\tilde{\varphi}$. □

Recall from chapter 4 that ω^Ξ represents an *integral* cohomology class so that from (8.4) we see that the energies of a harmonic map φ and a flag transform $\tilde{\varphi}$ differ by an integral amount. Indeed, if $\Xi: M \to F$ is a flag factor for φ and

$$\ldots \subset \mathbf{n}^i \subset \ldots \subset \mathbf{n} \subset \mathbf{q}^\Xi$$

is the central descending series for the nilradical bundle of $\mathbf{q}^\Xi$, from (4.24) we have

$$E(\varphi) - E(\tilde{\varphi}) = -4\int_M \omega^\Xi = 16\pi c_1(\bigoplus_i \mathbf{n}^i)[M] \in 16\pi\mathbf{Z} . \tag{9}$$

In fact, this result may be further refined: let F be a flag manifold corresponding to a parabolic subalgebra $\mathbf{q}_I$. By (4.20),

$$H_2(F,\mathbf{Z}) \cong \mathbf{Z}^{|I|}$$

with generators represented by the root spheres $\{\varphi_{\alpha_i} : i \in I\}$. If $\Xi: M \to F$ is a map of a closed Riemann surface, we define the *multi-degree* of Ξ to be the image under Ξ_* of the fundamental cycle of M viewed as an element of $\mathbf{Z}^{|I|}$. We can now apply (4.25) to see that if Ξ is a flag factor with multi-degree $(d_1,\ldots,d_{|I|})$, then

$$E(\varphi) - E(\tilde{\varphi}) = 16\pi \sum_{i \in I} \frac{d_i}{|\alpha_i|^2} . \tag{10}$$

Thus, since the length squared of any root divides that of the highest root, we conclude that

$$E(\varphi) - E(\tilde{\varphi}) \in 16\pi n_G \mathbf{Z} \tag{11}$$

where the integer n_G is the reciprocal of the length squared of the highest root of G with respect to the Killing metric. We list the values of n_G below:

G	n_G
$\mathbf{SU}(n)$	n
$\mathbf{Spin}(n)$, $n \geq 5$	$n-2$
$\mathbf{Sp}(n)$	$n+1$
$\mathbf{G}_2$	4
$\mathbf{F}_4$	9
$\mathbf{E}_6$	12
$\mathbf{E}_7$	18
$\mathbf{E}_8$	30

Flag transforms of maps into symmetric spaces

Let $\varphi: M \to N$ be a harmonic map into a symmetric G-space. Choosing a base-point in N, we obtain a totally geodesic immersion of N in G and, by composition, a harmonic map $M \to G$ also denoted by φ. From (1), we see that

$$\varphi^*\theta = -2\varphi^*\beta^N$$

so that the flag factor equations, which only involve $\varphi^*\theta$, do not depend on the choice of base-point. We also note that on $\underline{\mathbf{g}}$,

$$\nabla^{\varphi} = d - \mathrm{ad}\varphi^*\beta^N$$

so that the pull-back of the Levi-Civita connection on G coincides with the pull-back of the canonical connection on N.

If $\Xi: M \to F \subset \Gamma$ is a flag factor for φ, we obtain a flag transform $\tilde{\varphi}: M \to G$ and we enquire into the circumstances under which $\tilde{\varphi}$ also factors through a symmetric G-space. For this, we consider the fibre-wise symmetric decomposition

$$\underline{\mathbf{g}} = \varphi^{-1}[\mathbf{k}] \oplus \varphi^{-1}[\mathbf{p}] .$$

We say that $\Xi: M \to F \subset \Gamma$ *commutes* with Φ if $\Xi \in C^\infty(\varphi^{-1}[\mathbf{k}])$ i.e.

$$\Xi_x \in [\mathbf{k}]_{\varphi(x)}$$

for all $x \in M$. This is equivalent to

$$\mathbf{q}^{\Xi} = \mathbf{q}^{\Xi} \cap \varphi^{-1}[\mathbf{k}^{\mathbf{C}}] \oplus \mathbf{q}^{\Xi} \cap \varphi^{-1}[\mathbf{p}^{\mathbf{C}}]$$

and amounts to the assertion that the involutions of N at $\varphi(x)$ and $N(F)$ at $\pi(\Xi_x)$ commute.

Theorem 8.6. *Let G be of adjoint type (i.e. with trivial centre). Let $\varphi: M \to G$ factor through a symmetric G-space N as in (1.9) and let $\Xi: M \to F \subset \Gamma$ commute with φ. Then, after left translation by a constant, the product $\varphi\gamma(\Xi): M \to G$ also factors through the image of an totally geodesically immersed symmetric G-space.*

Proof. Fix a base-point $x_0 \in M$ and immerse N in G using $\varphi(x_0)$ as base-point. Let τ_x be the involution of G at $\varphi(x) \in N$. It is straight-forward to show that, for $g \in G$,

$$\varphi(x) g \varphi(x)^{-1} = \tau_{x_0}(\tau_x(g)) .$$

Since Ξ commutes with φ, we also have

$$\tau\Xi = \Xi \tag{12}$$

We define a map $\sigma: M \to \mathrm{Aut}(G)$ by

$$\sigma_x(g) = \gamma(\Xi_x) g^{\tau_x} \gamma(\Xi_x)^{-1}$$

for $g \in G$. Now τ_x and conjugation by $\gamma(\Xi_x)$ are involutions which commute by (12) so that each σ_x is an involution. However, the results of chapter 3 show that there are only a finite number of G-conjugacy classes of involutions in $\mathrm{Aut}(G)$, each of which is compact and connected so that we deduce that σ has image in a single conjugacy class.

Define $\tilde{\varphi}: M \to G$ by

$$\tilde{\varphi} = \gamma(\Xi_{x_0})^{-1} \varphi\gamma(\Xi) ,$$

which is a left translation of our usual flag transform. A short calculation gives

$$\tilde{\varphi}(x) g \tilde{\varphi}(x)^{-1} = \sigma_{x_0}(\sigma_x(g)) ,$$

for all $g \in G$. But, since σ_x is G-conjugate to σ_{x_0}, there is a $g_x \in G$ such that

$$\sigma_{x_0}(\sigma_x(g)) = (g_x^{\sigma_0} g_x^{-1}) g (g_x^{\sigma_0} g_x^{-1})^{-1}$$

so that, if the centre of G is trivial, we have

$$\tilde{\varphi}(x) = g_x^{\sigma_0} g_x^{-1} .$$

Thus, if N' denotes the symmetric G-space G/K^{σ_0}, we see that the image of $\tilde{\varphi}$ lies in that of N' under the totally geodesic immersion of (1.9). □

Remarks. (*a*) Of course, N and N' will not in general coincide. However, we do observe from the proof that N is inner if and only if N' is.

(*b*) The image of a symmetric space N in Ad(G) is in general only a locally symmetric space, in fact it is $\Sigma_N \backslash W$ (c.f. chapter 3). However, in our applications, M will be simply connected in which case we may lift $\tilde{\varphi}$ to get a harmonic map into N'.

C. Existence of flag factors and the factorisation theorem

The simplest harmonic maps $M \to G$ are the constant maps. Let us consider which harmonic maps arise by iterating the procedure of taking flag transforms starting with a constant map. We begin by examining the maps obtained by taking a single flag transform of a constant.

Theorem 8.7. *Let $\varphi : M \to G$ be a harmonic map. φ is a flag transform of a constant map if and only if, up to left translation by a constant, $\varphi = \gamma(\Xi)$ where $\Xi : M \to F \subset \Gamma$ is a super-horizontal holomorphic map.*

In this case

$$E(\varphi) = 16\pi \sum_{i \in I} \frac{d_i}{|\alpha_i|^2} , \tag{13}$$

where $(d_1, \ldots, d_{|I|})$ is the multi-degree of Ξ.

Proof. First, let $\psi : M \to G$ be a constant map. Then $\nabla^\psi = d$ and $\psi^*\theta$ vanishes so that the flag factor equations for $\Xi : M \to F \subset \Gamma$ reduce to

$$d\Xi(T^{0,1}M) \subset \mathbf{g}_1$$

or, equivalently,

$$\Xi^*\beta^F(T^{0,1}M) \subset \mathbf{g}_1 .$$

Thus Ξ is a flag factor for a constant map if and only if it is a super-horizontal anti-holomorphic map. In this case, of course, $-\Xi$ is super-horizontal and holomorphic. But observe that $\gamma(\Xi)^2$ has image in the centre of G which is finite so that $\gamma(\Xi)^2$ is constant. Therefore, $\gamma(\Xi)$ and $\gamma(-\Xi)$ differ by multiplication by a constant so that, up to such a multiplication, the flag transform of ψ by Ξ is $\gamma(-\Xi)$ and so of the required form.

Conversely, if $\Xi : M \to F \subset \Gamma$ is a super-horizontal holomorphic map and $\varphi = \gamma(\Xi)$, we claim that Ξ is a flag factor for φ. Given this claim, the flag transform of φ by Ξ is $\tilde{\varphi} = \gamma(\Xi)^2$ which is constant as we observed above. Thus, by (8.5), φ is the flag transform of the constant $\tilde{\varphi}$ by $-\Xi$ while (13) follows from (10) since here $E(\tilde{\varphi})$ vanishes. To prove the claim, let $P : \underline{\mathbf{g}}^{\mathbf{C}} \to [\mathbf{p}^{\mathbf{C}}] = \sum_{i \in 2\mathbf{Z}+1} \mathbf{g}_i$ be orthoprojection. Then, from (2),

$$\varphi^*\theta = \Xi^*\gamma^*\theta = -2P\Xi^*\beta^F ,$$

while, from (4),

$$\nabla^\varphi \Xi = [(1-P)\Xi^*\beta^F , \Xi] .$$

Since Ξ is super-horizontal and holomorphic, $(\Xi^*\beta^F)^{1,0}$ takes values in $\mathbf{g}_1$ so that $\nabla^\varphi\Xi$ vanishes and $\varphi^*\theta^{1,0}$ has image in $\mathbf{g}_1$. This establishes the claim and the theorem is now proved. □

The flag transforms of constant maps are the analogues for simple G of what Uhlenbeck calls 1-unitons in her study of harmonic maps into $\mathbf{U}(n)$ [72] and we see that they have a particularly simple structure: indeed, if $\varphi = \gamma(\Xi)$ with Ξ super-horizontal and holomorphic then φ factors through the inner symmetric space $N(F)$ and Ξ is a twistor lift for φ. Conversely, if $\varphi: M \to N$ is a harmonic map into an inner symmetric space with super-horizontal twistor lift Ξ, then, after a totally geodesic immersion into G via (1.9), φ is the flag transform of a constant by $-\xi$.

In particular, the preceding remarks apply to all harmonic maps $\varphi: S^2 \to S^{2n}$ or $\mathbf{C}P^n$. Indeed, from (6.20) we know that all such maps have super-horizontal twistor lifts. In this case, (13) imposes restrictions on the energy spectrum of such maps which will be explored in a later section.

Recall that in proving (6.20) essential use was made of the Birkhoff-Grothendieck theorem (6.8). We now refine our use of this theorem to show that, under favourable circumstances, we may obtain *all* harmonic maps $S^2 \to G$ by iterating the procedure of taking flag transforms of a constant.

Definition. A compact simple Lie group is of *type H* if it admits a Hermitian symmetric space as quotient, or, equivalently, if it has a simple root with coefficient one in the highest root (c.f. chapter three).

Remark. A glance at the extended Dynkin diagrams reveals that all compact Lie simple groups are of type H except $\mathbf{E}_8$, $\mathbf{F}_4$ and $\mathbf{G}_2$.

When G is of type H, Γ contains among its components copies of all the Hermitian symmetric G-spaces as height one flag manifolds. In this case, for M the Riemann sphere, we may ensure a supply of energy decreasing flag transformations:

Theorem 8.8. *Let G be of type H and $\varphi: S^2 \to G$ a non-constant harmonic map. Let $F \subset \Gamma$ be a height one flag manifold (equivalently, a Hermitian symmetric G-space). Then there is a flag factor $\Xi: S^2 \to F$ for φ with flag factor $\tilde{\varphi}$ such that*

$$E(\tilde{\varphi}) \leq E(\varphi) - 16\pi n_G \ .$$

Further, if φ factors through an inner symmetric G-space as in (1.9) then so does $\tilde{\varphi}$.

Proof. We equip $\mathbf{g}^{\mathbf{C}}$ with the Koszul-Malgrange holomorphic structure from ∇^φ. Let us reduce our problem to that of finding a suitable holomorphic sub-bundle of $\mathbf{g}^{\mathbf{C}}$. Indeed, F is a conjugacy class of height one parabolic sub-algebras so that, by (8.2), finding a flag factor $\Xi: S^2 \to F$ is equivalent to finding a holomorphic sub-bundle of parabolic sub-algebras $\mathbf{Q} \subset \mathbf{g}^{\mathbf{C}}$ with fibres in F such that $\varphi^*\theta^{1,0}$ takes values in $\mathbf{Q}$. Our flag factor is then just the canonical section of $\mathbf{Q}$. Finally, if $\tilde{\varphi}$ is the flag transform of φ determined by $\mathbf{Q}$ and $\mathbf{N}$ is the nilradical bundle of $\mathbf{Q}$, from (9) and the fact that $\mathbf{N}$ is abelian, we have

$$E(\tilde{\varphi}) = E(\varphi) - 16\pi \deg(\mathbf{N})$$

so that to establish the energy inequality, it suffices to find $\mathbf{Q}$ with $\deg(\mathbf{N})$ strictly positive.

For this, we fix a maximal torus $\mathbf{t} \subset \mathbf{g}$ and apply (6.8) and (6.9) to obtain a holomorphically trivial bundle of Cartan sub-algebras $\mathbf{a} = [\mathbf{t}^{\mathbf{C}}]$ with associated root bundle decomposition

$$\underline{\mathbf{g}}^{\mathbf{C}} = \mathbf{a}+\sum_{\alpha\in\Delta}\mathbf{g}^{\alpha}$$

together with a section $\xi_0\in H^0(\mathbf{a})$ for which $\deg(\mathbf{g}^\alpha)=\alpha(\xi_0)/\sqrt{-1}$. Here $\Delta\subset H^0(\mathbf{a}^*)$ is a copy of the root system $\Delta(\mathbf{g}^{\mathbf{C}},\mathbf{t}^{\mathbf{C}})$. As in chapter six, we choose a set of positive roots Δ^+ so that $\alpha(\xi_0)\geq 0$ for all $\alpha\in\Delta^+$ and let $\mathbf{b}$ be the corresponding bundle of Borel sub-algebras. Since φ is harmonic, $\varphi^*\theta^{1,0}:T^{1,0}S^2\to\underline{\mathbf{g}}^{\mathbf{C}}$ is holomorphic so that if φ is non-constant, we have

$$\sup\underline{\mathbf{g}}^{\mathbf{C}} \geq \inf T^{1,0}S^2 = 2\ , \tag{14}$$

while, since $\sup(\underline{\mathbf{g}}^{\mathbf{C}}/\mathbf{b})\leq 0$, we see that $\varphi^*\theta^{1,0}$ has image in $\mathbf{b}$.

There is a simple root $\alpha_i\in\Delta^+$ for which $\mathbf{q}_{\{i\}}$ lies in the conjugacy class F. We set

$$\mathbf{Q} = \sum_{n_{\{i\}}(\alpha)\geq 0}\mathbf{g}^{\alpha}$$

which is a holomorphic bundle of parabolic sub-algebras containing $\mathbf{b}$ and so containing the image of $\varphi^*\theta^{1,0}$. Each fibre of $\mathbf{Q}$ is conjugate to $\mathbf{q}_{\{i\}}$ so that the canonical section of $\mathbf{Q}$ is our flag factor $\Xi:S^2\to F$. As for deg(N), since $\mathbf{N}\subset\mathbf{b}$, $\mathbf{N}$ is a sum of root line bundles of non-negative degree. If deg(N) vanishes, then the highest root bundle (which is contained in N) has zero degree and then $\alpha(\xi_0)$ must vanish for all $\alpha\in\Delta$ contradicting (14). Thus deg(N) is positive and the main part of the proof is complete.

Finally suppose that φ factors through an inner symmetric space. We then have the symmetric decomposition

$$\underline{\mathbf{g}}^{\mathbf{C}} = [\mathbf{k}^{\mathbf{C}}]\oplus[\mathbf{p}^{\mathbf{C}}]$$

and we may choose our maximal torus $\mathbf{t}\subset\mathbf{k}$ so that $\mathbf{a}\subset[\mathbf{k}^{\mathbf{C}}]$. Then $\mathbf{b}$ and $\mathbf{Q}$ will be compatible with the symmetric decomposition and so by (8.6), $\tilde{\varphi}$ will also factor through an inner symmetric space. □

We may now iterate this procedure, reducing the energy by at least $16\pi n_G$ each time until a constant map is reached. Using (8.5) to invert the process, we conclude that any harmonic 2-sphere in a Lie group of type H can be obtained by repeated flag transformations of constants. Thus any such map may be expressed as a product of exponentials of flag factors and we have the following factorisation theorem.

Theorem 8.9. *Let $\varphi:S^2\to G$ be a harmonic map into a compact Lie group of type H. Then there is a number $N\in\mathbf{N}$, harmonic maps $\varphi_i:S^2\to G$, $0\leq i\leq N$, and maps $\Xi_i:S^2\to H_i\subset\Gamma$, $1\leq i\leq N$, into Hermitian symmetric spaces H_i such that*

(i) φ_0 is constant and $\varphi_N=\varphi$;

(ii) Ξ_i is a flag factor for φ_{i-1} and φ_i is the flag transform of φ_{i-1} by Ξ_i;

(iii) $N\leq E(\varphi)/16\pi n_G$.

Thus

$$\varphi = \varphi_0\exp\pi\Xi_1\ldots\exp\pi\Xi_N\ .$$

Further, if φ factors through an inner symmetric space then so does each φ_i.

Remarks. (i) The factorisation obtained here can be highly non-unique: one may freely choose the Hermitian symmetric spaces H_i at each step.

(ii) Since the H_i are height one flag manifolds the canonical fibration is essentially the identity map so that each $\gamma(\Xi_i)$ factors through a copy of H_i totally geodesically immersed in G. Thus φ is expressed as a product of maps into Hermitian symmetric spaces.

(iii) Theorem (8.9) is due to Uhlenbeck [72] in the case $G = \mathbf{SU}(n)$ but our method of proof is based on Valli's approach [73] to Uhlenbeck's theorem. In this case, the H_i are complex Grassmannians.

(iv) For $G = \mathbf{G}_2$, $\mathbf{F}_4$ or $\mathbf{E}_8$, theorem (8.9) remains open. Moreover, a new method would be required for the following reason. If $\varphi: S^2 \to \mathbf{G}_2/\mathbf{SO}(4)$ is a root sphere generating $\pi_2(\mathbf{G}_2/\mathbf{SO}(4))$ then, while φ is the flag transform of a constant, it can be shown that the corresponding bundle $\mathbf{N}$ must contain root bundles of negative degree and so $\mathbf{Q}$ cannot arise by our construction.

(v) Again, if a harmonic map φ factors through a non-inner symmetric space, our methods cannot provide flag transforms which also so factor. Indeed, if φ factors through $\mathbf{SU}(2n+1)/\mathbf{SO}(2n+1)$, then no height one parabolic subalgebra bundle containing $\mathbf{b}$ is compatible with the symmetric decomposition.

(vi) As a corollary to (8.9), we deduce that the energy of a harmonic map of a 2-sphere into a Lie group of type H is an integer multiple of $16\pi n_G$. In fact, this result is true for all simple Lie groups as was proved by Eells, Freed and Valli. We shall return to this topic in a later section.

D. An example

Let us illustrate these ideas by considering harmonic maps $S^2 \to \mathbf{SO}(2n)$ and flag factors taking values in $J^+(\mathbf{R}^{2n})$; the space of Hermitian almost complex structures on $\mathbf{R}^{2n}$ compatible with a fixed orientation.

If $j \in J^+(\mathbf{R}^{2n})$, then we have the type decomposition

$$(\mathbf{R}^{2n})^{\mathbf{C}} = V_j^+ \oplus V_j^-$$

where V_j^+ is the maximally isotropic $\sqrt{-1}$-eigenspace of j. Using the metric on $\mathbf{R}^{2n}$ to identify $\mathbf{so}(2n)$ with $\Lambda^2\mathbf{R}^{2n}$, we see that

$$\beta(T_j^{1,0}J^+(\mathbf{R}^{2n})) = \Lambda^2 V_j^+ ,$$

so that the (parabolic) stabiliser at j is

$$\Lambda^2 V_j^+ \oplus V_j^+ \otimes V_j^-$$

and the canonical element is $j/2 \in \mathbf{so}(2n)$. Thus $\gamma: J^+(\mathbf{R}^{2n}) \to \mathbf{SO}(2n)$ is given by

$$\gamma(j) = \exp \pi j/2 = j\,. \tag{15}$$

A map $\psi: M \to J^+(\mathbf{R}^{2n})$ is the same as a maximally isotropic sub-bundle ψ of $V = (\underline{\mathbf{R}}^{2n})^{\mathbf{C}}$. Indeed, ψ is just the $\sqrt{-1}$-eigenbundle of ψ. The corresponding bundle of parabolic subalgebras $\mathbf{q}^\psi \subset \underline{\mathbf{so}(2n)}$ is then given by

$$\mathbf{q}^\psi = \{\eta \in \underline{\mathbf{so}(2n)}: \eta\underline{\psi} \subset \underline{\psi}\} = \Lambda^2\underline{\psi} \oplus \underline{\psi} \otimes \overline{\underline{\psi}}\,.$$

Now let $\varphi: M \to \mathbf{SO}(2n)$ be a harmonic map. Then $\varphi^*\theta$ is an $\mathbf{so}(2n)$-valued 1-form and

$$\nabla^\varphi = d + \tfrac{1}{2}\varphi^*\theta$$

defines a connection (and hence holomorphic structure) on V which is compatible with that on $\underline{\mathbf{so}(2n)}$.

Recall that $\psi: M \to J^+(\mathbf{R}^{2n})$ is a flag factor for φ if and only if $\varphi^*\theta^{1,0}$ takes values in $\mathbf{q}^\psi$ and $\mathbf{q}^\psi$ is holomorphic. In our situation, the first condition means that

$$\varphi^*\theta^{1,0}\underline{\psi}\subset\kappa_M\otimes\underline{\psi}$$

while the second condition is just that $\underline{\psi}$ is holomorphic. Thus flag factors for φ with values in $J^+(\mathbf{R}^{2n})$ are the same as holomorphic maximally isotropic sub-bundles of V stable under the action of $\varphi^*\theta^{1,0}$. If $\underline{\psi}$ is such a bundle, then from (15), the corresponding flag transform $\tilde{\varphi}:M\to\mathbf{SO}(2n)$ is given by the point-wise product

$$\tilde{\varphi} = \varphi(x)\psi(x)$$

for $x\in M$. Suppose now that $\varphi:S^2\to\mathbf{SO}(2n)$ is non-constant and harmonic. We shall find an energy decreasing flag factor as in (8.8). We apply the Birkhoff-Grothendieck theorem (2.12) to V and argue as in (2.13) to produce a holomorphic maximally isotropic sub-bundle $\underline{\psi}\subset V$ with

$$\inf\underline{\psi}\geq 0, \qquad \sup(V/\underline{\psi})\leq 0\ .$$

Now $\varphi^*\theta^{1,0}$ restricts to give a holomorphic endomorphism $T^{1,0}S^2\otimes\underline{\psi}\to V$ which has image in $\underline{\psi}$ by (2.7) since

$$\inf(T^{1,0}S^2\otimes\underline{\psi})\geq\inf T^{1,0}S^2 = 2$$

and so ψ is a flag transform for φ.

As for the energies,

$$\deg(\mathrm{N}) = \deg(\Lambda^2\underline{\psi}) = (n-1)\deg(\underline{\psi})\ ,$$

so that if $\tilde{\varphi}$ is the flag transform of φ by ψ, we have

$$E(\tilde{\varphi}) = E(\varphi) - 16\pi(n-1)\deg(\underline{\psi})\ .$$

If $\deg(\underline{\psi})$ vanishes, then $\underline{\psi}$ and hence V is holomorphically trivial and $\varphi^*\theta^{1,0}$ vanishes. We therefore conclude that $\underline{\psi}$ has positive degree whence

$$E(\tilde{\varphi}) < E(\varphi)\ .$$

Indeed, we can say more: in this case $n_G = 2n-2$ so that we conclude from (4.25) that $\underline{\psi}$ has even degree. Thus

$$E(\tilde{\varphi}) \leq E(\varphi) - 16\pi(2n-2) = E(\varphi) - 16\pi n_G$$

as in (8.8).

Remarks. (i) A similar analysis can be given to prove (8.8) for the other classical groups. For example, when $G=\mathbf{SU}(n)$, an energy decreasing flag factor with values in a Grassmannian amounts to a holomorphic sub-bundle of $\underline{\mathbf{C}}^n$ of positive degree which is stable under $\varphi^*\theta^{1,0}$, in this case an $\mathbf{su}(n)$-valued 1-form. Such are easily found by the Birkhoff-Grothendieck theorem and, indeed, this is the approach that Valli took in his proof of Uhlenbeck's theorem.

(ii) This example also illustrates the difficulties involved in extending (8.8) to domains of higher genus. In that case, we must use the Harder-Narasimhan filtration of V but there is no guarantee that this will provide a maximally isotropic holomorphic $\underline{\psi}$ of positive slope (c.f. the similar difficulties encountered in chapter 2). Moreover, even if such a $\underline{\psi}$ existed, it need not be stable under $\varphi^*\theta^{1,0}$ unless the canonical bundle of M is non-negative or φ is suitably ramified. Thus, while partial results are available for domain a torus or for highly ramified φ we cannot prove a factorisation theorem by our technique. Indeed, we may encounter a $\underline{\mathbf{g}}^{\mathbf{C}}$ which is holomorphically non-trivial but semi-stable of slope zero and then no flag factors can be found by our methods. Similar limitations have been found by Wolfson [83] in his study of minimal surfaces in complex Grassmannians.

E. Applications

We now apply the above results to the study of the energy of harmonic 2-spheres in Lie groups and symmetric spaces. Since we have shown that the energy of such a map is quantised, we would expect to isolate various gap phenomena for harmonic 2-spheres. Such is indeed the case and we shall see that our setting provides a uniform context for some results of this kind already in the literature as well as some new ones.

We begin with some notational matters. Let G be a compact simple Lie group and h an invariant inner product on $\mathbf{g}$. Further, let N be a symmetric G-space. We may pull h back by the Maurer-Cartan forms to provide G and N with invariant metrics, also denoted by h. Finally, we denote by $n_G(h)$ the reciprocal of the length squared of the highest root of $\mathbf{g}$ with respect to h. All of this is consistent with our choice of metrics in the previous sections where we took h to be the negative of the Killing form with $n_G(h) = n_G$.

Since any invariant metric on $\mathbf{g}$ is a scalar multiple of the Killing metric, it follows from (8.9) and scaling that if $\varphi: S^2 \to G$ is a harmonic 2-sphere in a Lie group of type H then $E(\varphi)$ is an integer multiple of $16\pi n_G(h)$. In fact, as Eells-Freed and Valli have shown, this result as true even when G is not of type H. As a corollary, we have:

Proposition 8.10. *Let $\varphi: S^2 \to (G,h)$ be a harmonic map into a compact simple Lie group. If*

$$E(\varphi) < 16\pi n_G(h) ,$$

then φ is constant.

When G is of type H, we can do better:

Proposition 8.11. *Let G be of type H and $\varphi: S^2 \to (G,h)$ be harmonic. If*

$$E(\varphi) < 32\pi n_G(h) ,$$

then φ factors holomorphically through a Hermitian symmetric space.

Proof. If $E(\varphi) = 16\pi N_G(h)$ then the energy of any flag transform provided by theorem (8.8) must be zero. Thus φ is the flag transform of a constant by a flag factor with values in a Hermitian symmetric space so that the result follows from (8.7). □

Let us now consider harmonic 2-spheres in symmetric G-spaces. In this case, it is often possible to prove rather more refined results. We first observe that if $\varphi: S^2 \to (N,h)$ is a map into a symmetric G-space and $\Phi: N \to (G,h)$ is a totally geodesic immersion as in (1.9) then, by (1), we have

$$E(\Phi \circ \varphi) = 4E(\varphi) \tag{16}$$

so that if φ is harmonic then

$$E(\varphi) \in 4\pi n_G \mathbf{Z} . \tag{17}$$

Suppose now that N is an even-dimensional sphere S^{2n}. The only flag manifold that fibres canonically over N is the height two flag manifold $\mathbf{SO}(2n+1)/\mathbf{U}(n)$ which is the conjugacy class of a parabolic subalgebra $\mathbf{q}_{\{i\}}$ with α_i a short root. In chapter 6, we showed that any harmonic map $\varphi: S^2 \to S^{2n}$ has a superhorizontal holomorphic lift into $\mathbf{SO}(2n+1)/\mathbf{U}(n)$ whence, by (16) and (8.7), we conclude that

$$E(\varphi) = 4\pi \frac{d}{|\alpha_i|^2}$$

where d is the degree of the lift. Since α_i is short, this means that

$$E(\varphi) = \pi d n_G(h) .$$

Taking h to be the Killing metric gives $n_G(h) = 2n-1$, while an easy calculation shows that the canonical metric on S^{2n} differs from h by a factor of $4n-2$:

$$h = (4n-2) h_{\text{can}} ,$$

so that we obtain the following theorem of Barbosa [4]:

Theorem 8.12. *Let $\varphi : S^2 \to S^{2n}$ be a harmonic map into the unit 2n-sphere. Then*

$$E(\varphi) = 4\pi d \in 4\pi \mathbf{N} .$$

Now let us consider harmonic 2-spheres in Hermitian symmetric spaces. In this setting, the number $n_G(h)$ has a simple intrinsic interpretation provided by a lemma of Borel [7] (see also Nakagawa-Takagi [54]):

Lemma 8.13. *The maximum of the holomorphic sectional curvatures of a Hermitian symmetric G-space is $(n_G(h))^{-1}$.*

We now have

Theorem 8.14. *Let $\varphi : S^2 \to \mathbf{C}P^n(c)$ be a harmonic map into a complex projective space with a metric of constant holomorphic sectional curvature c. If φ is neither holomorphic or anti-holomorphic then*

$$E(\varphi) \geq \frac{4\pi}{c} \{3|\deg \varphi| + 4\} .$$

Proof. We shall equip $\mathbf{C}P^n$ with the Killing metric and prove that

$$E(\varphi) \geq 4\pi n_G \{3|\deg \varphi| + 4\} .$$

The result then follows from (8.13) by scaling the metric.

From chapter 6, we know that φ has a super-horizontal holomorphic twistor lift $\psi : S^2 \to F$ into a flag manifold F canonically fibred over $\mathbf{C}P^n$. By (4.26), such a flag manifold has height not exceeding two. Moreover the height one flag manifolds in this case are just copies of $\mathbf{C}P^n$ with canonical fibration the identity map so that, since φ is not $\pm$-holomorphic, we deduce that F is of height two. Thus F is of the form $\mathbf{SU}(n+1)/\mathbf{S}(\mathbf{U}(r) \times \mathbf{U}(1) \times \mathbf{U}(n-r))$.

From (8.7), we have

$$E(\varphi) = 4\pi(\deg \mathbf{n} + \deg [\mathbf{n}, \mathbf{n}])$$

where $\mathbf{n}$ is the nilradical bundle of $\mathbf{q}^{\psi}$. We realise $\mathbf{n}$ as follows (c.f. section E of chapter 4): we pull back the tautological bundles on F to get a orthogonal decomposition

$$\underline{\mathbf{C}}^{n+1} = E_- \oplus L \oplus E_+$$

where the summands have rank r, 1 and $n-r$ respectively. Then

$$\mathbf{n} = E_-^* \otimes L \oplus L^* \otimes E_+ \oplus E_-^* \otimes E_+ \;, \qquad [\mathbf{n}, \mathbf{n}] = E_-^* \otimes E_+$$

so that

$$\deg \mathbf{n} + \deg [\mathbf{n}, \mathbf{n}] = (n+1)(\deg E_+ + \deg (E_+ \oplus L)) = n_G(\deg E_+ + \deg (E_+ \oplus L)) \;.$$

Moreover, L is the pull-back by φ of the tautological bundle on $\mathbf{C}P^n$, the Chern class of which is easily seen to be the negative generator of $H^2(\mathbf{C}P^n, \mathbf{Z})$ so that

$$\deg \varphi = -\deg L \;.$$

Reversing the orientation of S^2 if necessary, we may assume that $\deg L$ is non-negative. Let β be the Maurer-Cartan form of $\mathbf{C}P^n$. Then, since ψ is holomorphic, we have

$$\varphi^* \beta^{1,0} L \subset T_{1,0}^* S^2 \otimes E_+ \;,$$

while, since φ is not anti-holomorphic, $\varphi^* \beta^{1,0}$ does not annihilate L and we conclude that

$$|\deg \varphi| + 2 \leq \sup E_+ .$$

Moreover, our lift ψ was constructed so that $\inf \mathbf{n} \geq 0$ while $L^* \otimes E_+$ is a parallel sub-bundle of $\mathbf{n}$ from which it follows that $\inf L^* \otimes E_+ \geq 0$ i.e.

$$0 \leq \deg \varphi \leq \inf E_+ \;.$$

From this we see that

$$\deg E_+ \geq \sup E \geq |\deg \varphi| + 2$$

whence

$$\deg \mathbf{n} + \deg [\mathbf{n}, \mathbf{n}] \geq n_G(3\,|\deg \varphi| + 4)$$

and the theorem is proved. □

Remark. In case $\deg \varphi = 0$, this result is due to Udagawa [70] who also showed that in this case the inequality is sharp. For $\mathbf{C}P^2$, the result is due to Catenacci-Martellini-Reina [26].

Finally, when N is a Hermitian symmetric space of non-constant holomorphic sectional curvatures, there is the following result of Udagawa[71]:

Theorem 8.15. *Let $\varphi : S^2 \to N$ be a harmonic map into a Hermitian symmetric space for which the maximum of the holomorphic sectional curvatures is c. Then if φ is neither holomorphic nor anti-holomorphic,*

$$E(\varphi) \geq \frac{4\pi}{c} \{ |\deg \varphi| + 2 \} \;.$$

Remark. For $\deg \varphi = 0$ and $N \neq \mathbf{C}P^n$, Udagawa shows that this inequality is sharp.

A proof of (8.15) from our point of view can be provided but to give the details here would take us rather far afield. We shall therefore only sketch the argument.

The idea is that G-flag manifolds may be holomorphically embedded in certain complex projective spaces as the projective orbits of a highest weight vector in a G-representation. In particular, to each Hermitian symmetric G-space N one may associate a certain fundamental representation of G with the following properties:

(i) N embeds holomorphically as the height weight orbit in the projectivised representation space;

(ii) this embedding induces an isomorphism of second cohomology groups.

For complex Grassmannians, this is just the Plücker embedding, while, for $\mathbf{SO}(2n)/\mathbf{U}(n)$ the representation is the spin representation realising this space as the space of projective pure spinors.

Let us now fix a Hermitian symmetric G-space N and corresponding fundamental representation V. Let $\varphi:S^2 \to G$ be harmonic. Then φ induces a connection and hence holomorphic structure on the trivial bundle $\underline{V}$ to which we apply the theorem of Birkhoff-Grothendieck. It can then be shown that if L is the line sub-bundle of highest degree in $\underline{V}$ then L determines a map $\psi:S^2 \to P(V)$ which has image in the highest weight orbit N. Moreover, ψ can be proved to be a flag factor for φ of degree $\deg L$.

Now suppose that φ itself factors through N. Then φ pulls back the tautological line bundle on $P(V)$ to give a parallel sub-bundle L^{φ} of $\underline{V}$ of degree $\deg\varphi$. If φ is not $\pm$-holomorphic, $\varphi^*\beta^{1,0}$ cannot annihilate L^{φ} so that we can conclude that

$$\deg L \geq |\deg\varphi| + 2 .$$

Since, by (11),

$$E(\varphi) \geq 4\pi n_G \deg L ,$$

theorem (8.15) follows.

We shall return to this approach to flag manifolds elsewhere.

F. Relationship with the loop group

The equations defining a flag factor seem somewhat mysterious but they can be motivated by introducing the infinite-dimensional Kähler manifold ΩG: the space of based loops on G. This approach, due to Uhlenbeck (*op. cit.*) is summarised below.

Let $\Omega G = L^1_2(S^1,1;G,e)$ denote the infinite-dimensional Banach manifold of based loops on G. Point-wise multiplication of loops endows ΩG with the structure of a Banach Lie group with Lie algebra $\Omega\mathbf{g} = L^1_2(S^1,1;\mathbf{g},e)$. In addition, ΩG has a left-invariant Kählerian complex structure given at the identity by

$$\Omega\mathbf{g}^{\mathbf{C}} = \{\sum_{n\in\mathbf{Z}} \xi_n(\lambda^{-n}-1) : \lambda\in S^1, \xi_n \in \mathbf{g}^{\mathbf{C}}\} \qquad \Omega^{1,0}\mathbf{g}^{\mathbf{C}} = \{\xi\in\Omega\mathbf{g}^{\mathbf{C}} : \xi_n = 0 \text{ for } n<0\} .$$

The Kähler form at the identity is

$$S(\xi,\eta) = \int_{S^1} (\xi',\eta)dt$$

and the Kähler metric is the (incomplete) $L^2_{\frac{1}{2}}$ metric. For more details on the geometry of ΩG the Reader is referred to the book of Pressley and Segal [58].

Now let $\varphi:S^2 \to G$ be a harmonic map and denote $\varphi^*\theta$ by A. Then the harmonicity of φ gives

$$d^*A = 0 , \tag{I}$$

while the pull-back of the Maurer-Cartan equations gives

$$dA + \tfrac{1}{2}[A\wedge A] = 0. \tag{II}$$

Define a family of G-connections on the trivial bundle $S^2\times G$ by

$$\nabla^{\lambda} = (\partial + \tfrac{1}{2}(1-\lambda^{-1})A^{1,0}, \bar\partial + \tfrac{1}{2}(1-\lambda)A^{0,1}) ,$$

for $\lambda \in S^1$. Then (I) and (II) are satisfied if and only if each ∇^λ is flat. In this case, since S^2 is simply-connected, each ∇^λ is gauge-equivalent to the trivial connection d so that we have a family of maps $\varphi_\lambda : S^2 \to G$, $\lambda \in S^1$, each defined up to left multiplication by a constant, such that

$$\varphi_\lambda{}^*\theta = \tfrac{1}{2}(1-\lambda^{-1})A^{1,0} + \tfrac{1}{2}(1-\lambda)A^{0,1} .$$

Choosing the constants so that $\varphi_1 \equiv e$, $\varphi_{-1} = \varphi$, we define a map $\Phi : S^2 \to \Omega G$ by

$$\Phi(x)(\lambda) = \varphi_\lambda(x) .$$

If Θ is the left Maurer-Cartan form of ΩG, then

$$\Phi^*\Theta(\lambda) = \tfrac{1}{2}(1-\lambda^{-1})A^{1,0} + \tfrac{1}{2}(1-\lambda)A^{0,1}$$

from which we conclude that Φ is holomorphic and *pseudo-horizontal* in the sense that

$$\mathrm{Im}\Phi^*\Theta \subset \mathbf{g}^{\mathbf{C}} \otimes \mathrm{span}\{\lambda^{-1}-1, \lambda-1\} .$$

Conversely, if $\Phi : S^2 \to \Omega G$ is holomorphic and pseudo-horizontal then $\Phi(1) : S^2 \to G$ is a harmonic map.

Remarks. (i) Holomorphic pseudo-horizontal maps are called *extended solutions* by Uhlenbeck.

(ii) These ideas may be put into the twistorial framework of the previous chapters since there is a left-invariant almost complex structure on ΩG which coincides with the Kähler structure on the pseudo-horizontal distribution and makes the map $\Omega G \to G$ given by evaluation at -1 into a twistor fibration. Thus the above construction provides twistor lifts for harmonic maps $S^2 \to G$. These ideas will be treated in detail elsewhere.

It is now convenient to assume that G is of adjoint type. In this case, we may embed the set of canonical elements into ΩG as the conjugacy classes of closed geodesics via $\iota : \Gamma \to \Omega G$ with $\iota(\xi)(e^{\sqrt{-1}t}) = \exp t\xi$. Each such conjugacy class is then a flag manifold, totally geodesically and holomorphically embedded in ΩG. We may now explain the significance of flag factors:

Proposition 8.16. *Let $\Phi : S^2 \to \Omega G$ be holomorphic and pseudo-horizontal with $\Phi(-1) = \varphi$ and let $\Xi : S^2 \to F \subset \Gamma$. Then the product*

$$\tilde{\Phi} = \Phi\iota(\Xi)$$

is holomorphic and pseudo-horizontal if and only if Ξ is a flag factor for φ and in this case $\tilde{\Phi}(-1)$ is the flag transform of φ by Ξ.

The (easy) proof is left to the Reader who has made it this far.

Remark. The assumption that G is of adjoint type is to ensure that $\exp 2\pi(\Gamma) = \{e\}$ and may be avoided by considering instead a suitable central extension of G.

References

1. Aithal, A.R.: Harmonic maps from S^2 to $G_{2,5}$. J. Lond. Math. Soc. **32**, 572–576 (1985)
2. Atiyah, M.F. and Bott, R.: The Yang-Mills equations over Riemann surfaces. Phil. Trans. R. Soc. Lond. A **308**, 523–615 (1982)
3. Bahy-El-Dien, A. and Wood, J.C.: The explicit construction of all harmonic two-spheres in quaternionic projective spaces. Leeds preprint, 1989
4. Barbosa, J.: On minimal immersions of S^2 into S^{2m}. Trans. Amer. Math. Soc. **210**, 75–106 (1975)
5. Baston, R.J. and Eastwood, M.G.: The Penrose transform: its interaction with representation theory. Oxford: Oxford Univ. Press 1989
6. Borel, A.: Kählerian coset spaces of semi-simple Lie groups. Proc. Nat. Acad. Sci. USA **40**, 1147–151 (1954)
7. Borel, A.: On the curvature tensor of the Hermitian symmetric manifolds. Ann. Math. **71**, 508–521 (1960)
8. Borel, A. and Hirzebruch, F.: Characteristic classes and homogeneous spaces, I. Amer. J. Math. **80**, 459–538 (1958)
9. Borel, A. and de Siebenthal, J.: Les sous-groupes fermés de rang maximum des groupes de Lie clos. Comm. Math. Helv. **23**, 200–221 (1949)
10. Bryant, R.L.: Conformal and minimal immersions of compact surfaces into the 4-sphere. J. Diff. Geom. **17**, 455–473 (1982)
11. Bryant, R.L.: Lie groups and twistor spaces. Duke Math. J. **52**, 223–261 (1985)
12. Burns, D.: Harmonic maps from $\mathbf{C}P^1$ to $\mathbf{C}P^n$. In: Proc. Tulane Conf.. R. Knill (ed.). Lect. Notes in Math. 1263. Berlin, Heidelberg, New York: Springer 1982
13. Burns, D., Burstall, F.E., de Bartolomeis, P., and Rawnsley, J.H.: Stability of harmonic maps of Kaehler manifolds. J. Diff. Geom. **30**, 579–594 (1989)
14. Burstall, F.E.: Twistor fibrations of flag manifolds and harmonic maps of a 2-sphere into a Grassmannian. In: Differential Geometry. L. Cordero (ed.). Pitman Research Notes in Math. 131. London: Pitman 1985
15. Burstall, F.E.: A twistor description of harmonic maps of a 2-sphere into a Grassmannian. Math. Ann. **274**, 61–74 (1986)
16. Burstall, F.E.: Twistor methods for harmonic maps. In: Differential Geometry. V.L. Hansen (ed.). Lect. Notes in Math. 1263. Berlin, Heidelberg, New York: Springer 1987
17. Burstall, F.E.: Recent developments in twistor methods for harmonic maps. In: Harmonic mappings, twistors and σ-models. P. Gauduchon (ed.). Singapore: World Scientific 1988
18. Burstall, F.E. and Rawnsley, J.H.: Sphères harmoniques dans les groupes de Lie compacts et courbes holomorphes dans les espaces homogènes. C. R. Acad. Sci. Paris **302**, 709–712 (1986)

19. Burstall, F.E. and Rawnsley, J.H.: Twistors, flags and symmetric spaces. Bath-Warwick preprint, 1989

20. Burstall, F.E., Rawnsley, J.H., and Salamon, S.M.: Stable harmonic 2-spheres in symmetric spaces. Bull. Amer. Math. Soc. **16**, 274–278 (1987)

21. Burstall, F.E. and Salamon, S.M.: Tournaments, flags and harmonic maps. Math. Ann. **277**, 249–265 (1987)

22. Burstall, F.E. and Wood, J.C.: The construction of harmonic maps into complex Grassmannians. J. Diff. Geom. **23**, 255–297 (1986)

23. Calabi, E.: Minimal immersions of surfaces in Euclidean spheres. J. Diff. Geom. **1**, 111–125 (1967)

24. Calabi, E.: Quelques applications de l'analyse complexe aux surfaces d'aire minima. In: Topics in Complex Manifolds. Université de Montréal 1967

25. Cartan, E.: Les groupes réels simple, finis et continues. Ann. Ec. Norm. Sup. **31**, 263–355 (1914)

26. Catenacci, R., Martellini, M., and Reina, C.: On the classical energy spectrum of $\mathbf{C}P^2$ models. Phys. Lett. B **115**, 461–462 (1982)

27. Cheeger, J. and Ebin, D.: Comparison Theorems in Riemannian Geometry. Amsterdam: North Holland 1975

28. Chern, S.S. and Wolfson, J.: Harmonic maps of the two-sphere into a complex Grassmann manifold, II. Ann. of Math. **125**, 301–335 (1987)

29. Din, A. M. and Zakrzewski, W. J.: General classical solutions in the $\mathbf{C}P^{n-1}$ model. Nucl. Phys. B. **174**, 397–406 (1980)

30. Dubois-Violette, M.: Structures complexes au-dessus des variétés. In: Mathématiques et Physique. Progress in Math., 37. Boston: Birkhäuser 1983

31. Eells, J. and Lemaire, L.: A report on harmonic maps. Bull. Lond. Math. Soc. **10**, 1–68 (1978)

32. Eells, J. and Lemaire, L.: Selected topics in harmonic maps. C.B.M.S Regional Conf. Ser. in Math. Providence, RI: Amer. Math. Soc. 1983

33. Eells, J. and Salamon, S.M.: Twistorial construction of harmonic maps of surfaces into four manifolds. Ann. Scuola Norm. Sup. Pisa **12**, 589–640 (1985)

34. Eells, J. and Wood, J.C.: Harmonic maps from surfaces into projective spaces. Adv. in Math. **49**, 217–263 (1983)

35. Erdem, S. and Wood, J.C.: On the construction of harmonic maps into a Grassmannian. J. Lond. Math. Soc. **28**, 161–174 (1983)

36. Glaser, V. and Stora, R.: Regular solutions of the $\mathbf{C}P^n$ model and further generalisations. CERN preprint, 1980

37. Gordon, W.B.: Convex functions and harmonic maps. Proc. Amer. Math. Soc. **33**, 433-437 (1972)

38. Griffiths, P.A.: Periods of integrals on algebraic manifolds, III. Publ. Math. I.H.E.S **38**, 125–180 (1970)

39. Gromov, M.: Pseudo holomorphic curves in symplectic manifolds. Invent. Math. **82**, 307–347 (1985)

40. Grothendieck, A.: Sur la classification des fibrés holomorphes sur la sphère de Riemann. Am. J. Math. **79**, 121–138 (1957)

41. Harder, G. and Narasimhan, M.S.: On the cohomology groups of moduli spaces of vector bundles over curves. Math. Ann. **212**, 215–248 (1975)

42. Helgason, S.: Differential Geometry, Lie Groups, and Symmetric Spaces. New York, San Francisco, London: Academic Press 1978

43. Hitchin, N.J.: Kählerian twistor spaces. Proc. Lond. Math. Soc. **43**, 133–150 (1981)

44. Horváthy, P.A. and Rawnsley, J.H.: Monopole charges for arbitrary compact gauge groups and Higgs fields in in any representation. Comm. Math. Phys. **99**, 517–540 (1985)

45. Howard, R. and Wei, W.: Nonexistence of stable harmonic maps to and from certain homogeneous spaces and submanifolds of Euclidean spaces. Trans. Amer. Math. Soc. **294**, 319–331 (1986)

46. Humphreys, J.E.: Introduction to Lie algebras and representation theory. New York, Heidelberg, London: Springer 1972

47. Jiménez, J.A.: Riemannian 4-symmetric spaces. Trans. Amer. Math. Soc. **306**, 715–734 (1988)

48. Koszul, J.L and Malgrange, B.: Sur certaines structures fibrées complexes. Arch. Math. **9**, 102–109 (1958)

49. Lang, S.: On quasi-algebraic closure. Ann. Math. **55**, 373–390 (1952)

50. Ledger, A.J.: Espace de Riemann symmetriques generalisés. C. R. Acad. Sci. Paris **264**, 947–948 (1967)

51. Lichnerowicz, A.: Applications harmoniques et variétés kählériennes. Sympos. Math. **3**, 341–402 (1970)

52. Micallef, M. and Moore, J.D.: Minimal two-spheres and the topology of manifolds with positive curvature on totally isotropic two-planes. Ann. of Math. **127**, 199–227 (1988)

53. Murakami, S.: Sur la classification des algèbres de Lie réelles et simples. Osaka J. Math. **2**, 291–307 (1965)

54. Nakagawa, H. and Takagi, R.: On locally symmetric Kaehler submanifolds in a complex projective space. J. Math. Soc. Japan **28**, 638–667 (1976)

55. O'Brian, N.R. and Rawnsley, J.H.: Twistor spaces. Ann. Glob. Anal. and Geom. **3**, 29–58 (1985)

56. Ohnita, Y.: Stability of harmonic maps and standard minimal immersions. Tôhuku Math. J. **38**, 259–267 (1986)

57. Pluzhnikov, A.I.: On the minimum of the Dirichlet functional. Sov. Math. Dokl. **34**, 281–284 (1987)

58. Pressley, A.N. and Segal, G.: Loop Groups. Oxford Math. Monographs. Oxford: Clarendon Press 1986

59. Ramanathan, J.: Harmonic maps from S^2 to $G_{2,4}$. J. Diff. Geom. **19**, 207–219 (1984)

60. Rawnsley, J.H.: Noether's theorem for harmonic maps. In: Diff. Geom. Methods in Math. Phys.. S. Sternberg (ed.). Dortrecht, Boston, London: Reidel 1984

61. Rawnsley, J.H.: f-structures, f-twistor spaces and harmonic maps. In: Geometry Seminar L. Bianchi.. Lect. Notes Math. 1164. Berlin, Heidelberg, New York: Springer 1986

62. Salamon, S.M.: Harmonic and holomorphic maps. In: Geometry Seminar L. Bianchi.. Lect. Notes Math. 1164. Berlin, Heidelberg, New York: Springer 1986

63. Sampson, J.H.: Some properties and applications of harmonic mappings. Ann. Sci. Ecole Norm. Sup. **11**, 211–228 (1978)

64. Schmid, W.: Homogeneous complex manifolds and representations of semisimple Lie groups. Berkeley thesis, 1967

65. Serre, J.-P.: Répresentations lineares et espaces homogènes complexes. Séminaire Bourbaki **100**(1959)

66. Siu, Y.T.: Curvature characterization of the hyperquadrics. Duke Math. J. **47**, 641–654 (1980)

67. Siu, Y.T. and Yau, S.T.: Compact Kähler manifolds of positive bisectional curvature. Invent. Math. **59**, 189–204 (1980)

68. Smith, R.T.: The second variation formula for harmonic mappings. Proc. Amer. Math. Soc. **47**, 229–236 (1975)

69. Takeuchi, M.: On the fundamental group and the group of isometries of a symmetric space. J. Fac. Sci. Univ. Tokyo **10**, 88–123 (1964)

70. Udagawa, S.: Pluriharmonic maps and minimal immersions of Kähler manifolds. J. Lond. Math. Soc. **37**, 375–384 (1988)

71. Udagawa, S.: Holomorphicity of certain stable harmonic maps and minimal immersions. Proc. Lond. Math. Soc. **57**, 577–598 (1988)

72. Uhlenbeck, K.: Harmonic maps into Lie groups (classical solutions of the chiral model). J. Diff. Geom. **30**, 1–50 (1989)

73. Valli, G.: On the energy spectrum of harmonic 2-spheres in unitary groups. Topology **277**, 129–136 (1988)

74. Varadarajan, V.S.: Lie Groups, Lie Algebras and their Representations. Berlin, Heidelberg, New York, Tokyo: Springer Verlag 1984

75. Wallach, N.: Harmonic analysis in homogeneous spaces. New York: Marcel Dekker 1973

76. Warner, G.: Harmonic analysis on semisimple Lie groups, I. New York, Heidelberg, London: Springer 1972

77. Wells, R.O.: Differential Analysis on Complex Manifolds. Berlin, Heidelberg, New York: Springer 1980

78. Wells, R.O. and Wolf, J.A.: Poincaré series and automorphic cohomology on flag domains. Ann. Math. **105**, 397–448 (1977)

79. Wolf, J.A.: Spaces of constant curvature. New York: McGraw-Hill 1967

80. Wolf, J.A.: The action of a real semisimple Lie group on a complex flag manifold. I: Orbit structure and holomorphic arc components. Bull. Amer. Math. Soc. **75**, 1121–1237 (1969)

81. Wolf, J.A. and Gray, A.: Homogeneous spaces defined by Lie group automorphisms, I. J. Diff. Geom. **2**, 77–114 (1968)

82. Wolfson, J.: Harmonic maps of the two-sphere into the complex hyperquadric. J. Diff. Geom. **24**, 141–152 (1986)

83. Wolfson, J.C.: Harmonic sequences and harmonic maps of surfaces into complex Grassman manifolds. J. Diff. Geom. **27**, 161–178 (1988)

84. Wood, J.C.: Twistor constructions for harmonic maps. In: Differential Geometry and Differential Equations. C.H. Gu (ed.). Lect. Notes in Math. 1255. Berlin, Heidelberg, New York: Springer 1988

85. Wood, J.C.: Explicit construction and parametrisation of harmonic two-spheres in the unitary group. Proc. Lond. Math. Soc. **58**, 608–624 (1989)

86. Zakrzewski, W.J.: Classical solutions of two-dimensional Grassmannian models. J. Geom. Phys. **1**, 39–63 (1984)

87. Zhong, J.-Q.: The degree of strong degeneracy of the bisectional curvature of exceptional bounded symmetric domains. In: Proc. Internat. Conf. Several Complex Variables Hangzhore. Boston: Birkhäuser 1984

Index

LECTURE NOTES IN MATHEMATICS

Edited by A. Dold, B. Eckmann and F. Takens

Some general remarks on the publication of monographs and seminars

In what follows all references to monographs, are applicable also to multiauthorship volumes such as seminar notes.

§1. Lecture Notes aim to report new developments - quickly, informally, and at a high level. Monograph manuscripts should be reasonably self-contained and rounded off. Thus they may, and often will, present not only results of the author but also related work by other people. Furthermore, the manuscripts should provide sufficient motivation, examples and applications. This clearly distinguishes Lecture Notes manuscripts from journal articles which normally are very concise. Articles intended for a journal but too long to be accepted by most journals, usually do not have this "lecture notes" character. For similar reasons it is unusual for Ph.D. theses to be accepted for the Lecture Notes series.

Experience has shown that English language manuscripts achieve a much wider distribution.

§2. Manuscripts or plans for Lecture Notes volumes should be submitted (preferably in duplicate) either to one of the series editors or to Springer- Verlag, Heidelberg. These proposals are then refereed. A final decision concerning publication can only be made on the basis of the complete manuscripts, but a preliminary decision can usually be based on partial information: a fairly detailed outline describing the planned contents of each chapter, and an indication of the estimated length, a bibliography, and one or two sample chapters - or a first draft of the manuscript. The editors will try to make the preliminary decision as definite as they can on the basis of the available information. We generally advise authors not to prepare the final master copy of their manuscript (cf. §4) beforehand.

§3. Final manuscripts should contain at least 100 pages of mathematical text and should include

- a table of contents;
- an informative introduction, perhaps with some historical remarks: it should be accessible to a reader not particularly familiar with the topic treated;
- a subject index: this is almost always genuinely helpful for the reader.

§4. Lecture Notes are printed by photo-offset from the master-copy delivered in camera-ready form by the authors. Springer-Verlag provides technical instructions for the preparation of manuscripts, for typewritten manuscripts special stationery, with the prescribed typing area outlined, is available on request. Careful preparation of the manuscripts will help keep production time short and ensure satisfactory appearance of the finished book. For manuscripts typed or typeset according to our instructions, Springer-Verlag will, if necessary, contribute towards the preparation costs at a fixed rate.

The actual production of a Lecture Notes volume takes 6-8 weeks.

§5. Authors receive a total of 50 free copies of their volume, but no royalties. They are entitled to purchase further copies of their book for their personal use at a discount of 33.3 %, other Springer mathematics books at a discount of 20 % directly from Springer-Verlag.

Commitment to publish is made by letter of intent rather than by signing a formal contract. Springer-Verlag secures the copyright for each volume.

<u>Addresses:</u>

Professor A. Dold, Mathematisches Institut, Universität Heidelberg,
Im Neuenheimer Feld 288, 6900 Heidelberg, Federal Republic of Germany

Professor B. Eckmann, Mathematik, ETH-Zentrum
8092 Zürich, Switzerland

Prof. F. Takens, Mathematisch Instituut, Rijksuniversiteit Groningen,
Postbus 800, 9700 AV Groningen, The Netherlands

Springer-Verlag, Mathematics Editorial, Tiergartenstr. 17,
6900 Heidelberg, Federal Republic of Germany, Tel.: (06221) 487-410

Springer-Verlag, Mathematics Editorial, 175 Fifth Avenue,
New York, New York 10010, USA, Tel.: (212) 460-1596

Vol. 1259: F. Cano Torres, Desingularization Strategies for Three-Dimensional Vector Fields. IX, 189 pages. 1987.

Vol. 1260: N.H. Pavel, Nonlinear Evolution Operators and Semi-groups. VI, 285 pages. 1987.

Vol. 1261: H. Abels, Finite Presentability of S-Arithmetic Groups. Compact Presentability of Solvable Groups. VI, 178 pages. 1987.

Vol. 1262: E. Hlawka (Hrsg.), Zahlentheoretische Analysis II. Seminar, 1984–86. V, 158 Seiten. 1987.

Vol. 1263: V.L. Hansen (Ed.), Differential Geometry. Proceedings, 1985. XI, 288 pages. 1987.

Vol. 1264: Wu Wen-tsün, Rational Homotopy Type. VIII, 219 pages. 1987.

Vol. 1265: W. Van Assche, Asymptotics for Orthogonal Polynomials. VI, 201 pages. 1987.

Vol. 1266: F. Ghione, C. Peskine, E. Sernesi (Eds.), Space Curves. Proceedings, 1985. VI, 272 pages. 1987.

Vol. 1267: J. Lindenstrauss, V.D. Milman (Eds.), Geometrical Aspects of Functional Analysis. Seminar. VII, 212 pages. 1987.

Vol. 1268: S.G. Krantz (Ed.), Complex Analysis. Seminar, 1986. VII, 195 pages. 1987.

Vol. 1269: M. Shiota, Nash Manifolds. VI, 223 pages. 1987.

Vol. 1270: C. Carasso, P.-A. Raviart, D. Serre (Eds.), Nonlinear Hyperbolic Problems. Proceedings, 1986. XV, 341 pages. 1987.

Vol. 1271: A.M. Cohen, W.H. Hesselink, W.L.J. van der Kallen, J.R. Strooker (Eds.), Algebraic Groups Utrecht 1986. Proceedings. XII, 284 pages. 1987.

Vol. 1272: M.S. Livšic, L.L. Waksman, Commuting Nonselfadjoint Operators in Hilbert Space. III, 115 pages. 1987.

Vol. 1273: G.-M. Greuel, G. Trautmann (Eds.), Singularities, Representation of Algebras, and Vector Bundles. Proceedings, 1985. XIV, 383 pages. 1987.

Vol. 1274: N. C. Phillips, Equivariant K-Theory and Freeness of Group Actions on C*-Algebras. VIII, 371 pages. 1987.

Vol. 1275: C.A. Berenstein (Ed.), Complex Analysis I. Proceedings, 1985–86. XV, 331 pages. 1987.

Vol. 1276: C.A. Berenstein (Ed.), Complex Analysis II. Proceedings, 1985–86. IX, 320 pages. 1987.

Vol. 1277: C.A. Berenstein (Ed.), Complex Analysis III. Proceedings, 1985–86. X, 350 pages. 1987.

Vol. 1278: S.S. Koh (Ed.), Invariant Theory. Proceedings, 1985. V, 102 pages. 1987.

Vol. 1279: D. Ieşan, Saint-Venant's Problem. VIII, 162 Seiten. 1987.

Vol. 1280: E. Neher, Jordan Triple Systems by the Grid Approach. XII, 193 pages. 1987.

Vol. 1281: O.H. Kegel, F. Menegazzo, G. Zacher (Eds.), Group Theory. Proceedings, 1986. VII, 179 pages. 1987.

Vol. 1282: D.E. Handelman, Positive Polynomials, Convex Integral Polytopes, and a Random Walk Problem. XI, 136 pages. 1987.

Vol. 1283: S. Mardešić, J. Segal (Eds.), Geometric Topology and Shape Theory. Proceedings, 1986. V, 261 pages. 1987.

Vol. 1284: B.H. Matzat, Konstruktive Galoistheorie. X, 286 pages. 1987.

Vol. 1285: I.W. Knowles, Y. Saitō (Eds.), Differential Equations and Mathematical Physics. Proceedings, 1986. XVI, 499 pages. 1987.

Vol. 1286: H.R. Miller, D.C. Ravenel (Eds.), Algebraic Topology. Proceedings, 1986. VII, 341 pages. 1987.

Vol. 1287: E.B. Saff (Ed.), Approximation Theory, Tampa. Proceedings, 1985–1986. V, 228 pages. 1987.

Vol. 1288: Yu. L. Rodin, Generalized Analytic Functions on Riemann Surfaces. V, 128 pages, 1987.

Vol. 1289: Yu. I. Manin (Ed.), K-Theory, Arithmetic and Geometry. Seminar, 1984–1986. V, 399 pages. 1987.

Vol. 1290: G. Wüstholz (Ed.), Diophantine Approximation and Transcendence Theory. Seminar, 1985. V, 243 pages. 1987.

Vol. 1291: C. Mœglin, M.-F. Vignéras, J.-L. Waldspurger, Correspondances de Howe sur un Corps p-adique. VII, 163 pages. 1987

Vol. 1292: J.T. Baldwin (Ed.), Classification Theory. Proceedings, 1985. VI, 500 pages. 1987.

Vol. 1293: W. Ebeling, The Monodromy Groups of Isolated Singularities of Complete Intersections. XIV, 153 pages. 1987.

Vol. 1294: M. Queffélec, Substitution Dynamical Systems – Spectral Analysis. XIII, 240 pages. 1987.

Vol. 1295: P. Lelong, P. Dolbeault, H. Skoda (Réd.), Séminaire d'Analyse P. Lelong – P. Dolbeault – H. Skoda. Seminar, 1985/1986. VII, 283 pages. 1987.

Vol. 1296: M.-P. Malliavin (Ed.), Séminaire d'Algèbre Paul Dubreil et Marie-Paule Malliavin. Proceedings, 1986. IV, 324 pages. 1987.

Vol. 1297: Zhu Y.-l., Guo B.-y. (Eds.), Numerical Methods for Partial Differential Equations. Proceedings. XI, 244 pages. 1987.

Vol. 1298: J. Aguadé, R. Kane (Eds.), Algebraic Topology, Barcelona 1986. Proceedings. X, 255 pages. 1987.

Vol. 1299: S. Watanabe, Yu.V. Prokhorov (Eds.), Probability Theory and Mathematical Statistics. Proceedings, 1986. VIII, 589 pages. 1988.

Vol. 1300: G.B. Seligman, Constructions of Lie Algebras and their Modules. VI, 190 pages. 1988.

Vol. 1301: N. Schappacher, Periods of Hecke Characters. XV, 160 pages. 1988.

Vol. 1302: M. Cwikel, J. Peetre, Y. Sagher, H. Wallin (Eds.), Function Spaces and Applications. Proceedings, 1986. VI, 445 pages. 1988.

Vol. 1303: L. Accardi, W. von Waldenfels (Eds.), Quantum Probability and Applications III. Proceedings, 1987. VI, 373 pages. 1988.

Vol. 1304: F.Q. Gouvêa, Arithmetic of *p*-adic Modular Forms. VIII, 121 pages. 1988.

Vol. 1305: D.S. Lubinsky, E.B. Saff, Strong Asymptotics for Extremal Polynomials Associated with Weights on $\mathbb{R}$. VII, 153 pages. 1988.

Vol. 1306: S.S. Chern (Ed.), Partial Differential Equations. Proceedings, 1986. VI, 294 pages. 1988.

Vol. 1307: T. Murai, A Real Variable Method for the Cauchy Transform, and Analytic Capacity. VIII, 133 pages. 1988.

Vol. 1308: P. Imkeller, Two-Parameter Martingales and Their Quadratic Variation. IV, 177 pages. 1988.

Vol. 1309: B. Fiedler, Global Bifurcation of Periodic Solutions with Symmetry. VIII, 144 pages. 1988.

Vol. 1310: O.A. Laudal, G. Pfister, Local Moduli and Singularities. V, 117 pages. 1988.

Vol. 1311: A. Holme, R. Speiser (Eds.), Algebraic Geometry, Sundance 1986. Proceedings. VI, 320 pages. 1988.

Vol. 1312: N.A. Shirokov, Analytic Functions Smooth up to the Boundary. III, 213 pages. 1988.

Vol. 1313: F. Colonius, Optimal Periodic Control. VI, 177 pages. 1988.

Vol. 1314: A. Futaki, Kähler-Einstein Metrics and Integral Invariants. IV, 140 pages. 1988.

Vol. 1315: R.A. McCoy, I. Ntantu, Topological Properties of Spaces of Continuous Functions. IV, 124 pages. 1988.

Vol. 1316: H. Korezlioglu, A.S. Ustunel (Eds.), Stochastic Analysis and Related Topics. Proceedings, 1986. V, 371 pages. 1988.

Vol. 1317: J. Lindenstrauss, V.D. Milman (Eds.), Geometric Aspects of Functional Analysis. Seminar, 1986–87. VII, 289 pages. 1988.

Vol. 1318: Y. Felix (Ed.), Algebraic Topology – Rational Homotopy. Proceedings, 1986. VIII, 245 pages. 1988

Vol. 1319: M. Vuorinen, Conformal Geometry and Quasiregular Mappings. XIX, 209 pages. 1988.

Vol. 1320: H. Jürgensen, G. Lallement, H.J. Weinert (Eds.), Semigroups, Theory and Applications. Proceedings, 1986. X, 416 pages. 1988.

Vol. 1321: J. Azéma, P.A. Meyer, M. Yor (Eds.), Séminaire de Probabilités XXII. Proceedings. IV, 600 pages. 1988.

Vol. 1322: M. Métivier, S. Watanabe (Eds.), Stochastic Analysis. Proceedings, 1987. VII, 197 pages. 1988.

Vol. 1323: D.R. Anderson, H.J. Munkholm, Boundedly Controlled Topology. XII, 309 pages. 1988.

Vol. 1324: F. Cardoso, D.G. de Figueiredo, R. Iório, O. Lopes (Eds.), Partial Differential Equations. Proceedings, 1986. VIII, 433 pages. 1988.

Vol. 1325: A. Truman, I.M. Davies (Eds.), Stochastic Mechanics and Stochastic Processes. Proceedings, 1986. V, 220 pages. 1988.

Vol. 1326: P.S. Landweber (Ed.), Elliptic Curves and Modular Forms in Algebraic Topology. Proceedings, 1986. V, 224 pages. 1988.

Vol. 1327: W. Bruns, U. Vetter, Determinantal Rings. VII,236 pages. 1988.

Vol. 1328: J.L. Bueso, P. Jara, B. Torrecillas (Eds.), Ring Theory. Proceedings, 1986. IX, 331 pages. 1988.

Vol. 1329: M. Alfaro, J.S. Dehesa, F.J. Marcellan, J.L. Rubio de Francia, J. Vinuesa (Eds.): Orthogonal Polynomials and their Applications. Proceedings, 1986. XV, 334 pages. 1988.

Vol. 1330: A. Ambrosetti, F. Gori, R. Lucchetti (Eds.), Mathematical Economics. Montecatini Terme 1986. Seminar. VII, 137 pages. 1988.

Vol. 1331: R. Bamón, R. Labarca, J. Palis Jr. (Eds.), Dynamical Systems, Valparaiso 1986. Proceedings. VI, 250 pages. 1988.

Vol. 1332: E. Odell, H. Rosenthal (Eds.), Functional Analysis. Proceedings, 1986–87. V, 202 pages. 1988.

Vol. 1333: A.S. Kechris, D.A. Martin, J.R. Steel (Eds.), Cabal Seminar 81–85. Proceedings, 1981–85. V, 224 pages. 1988.

Vol. 1334: Yu.G. Borisovich, Yu. E. Gliklikh (Eds.), Global Analysis – Studies and Applications III. V, 331 pages. 1988.

Vol. 1335: F. Guillén, V. Navarro Aznar, P. Pascual-Gainza, F. Puerta, Hyperrésolutions cubiques et descente cohomologique. XII, 192 pages. 1988.

Vol. 1336: B. Helffer, Semi-Classical Analysis for the Schrödinger Operator and Applications. V, 107 pages. 1988.

Vol. 1337: E. Sernesi (Ed.), Theory of Moduli. Seminar, 1985. VIII, 232 pages. 1988.

Vol. 1338: A.B. Mingarelli, S.G. Halvorsen, Non-Oscillation Domains of Differential Equations with Two Parameters. XI, 109 pages. 1988.

Vol. 1339: T. Sunada (Ed.), Geometry and Analysis of Manifolds. Procedings, 1987. IX, 277 pages. 1988.

Vol. 1340: S. Hildebrandt, D.S. Kinderlehrer, M. Miranda (Eds.), Calculus of Variations and Partial Differential Equations. Proceedings, 1986. IX, 301 pages. 1988.

Vol. 1341: M. Dauge, Elliptic Boundary Value Problems on Corner Domains. VIII, 259 pages. 1988.

Vol. 1342: J.C. Alexander (Ed.), Dynamical Systems. Proceedings, 1986–87. VIII, 726 pages. 1988.

Vol. 1343: H. Ulrich, Fixed Point Theory of Parametrized Equivariant Maps. VII, 147 pages. 1988.

Vol. 1344: J. Král, J. Lukeš, J. Netuka, J. Veselý (Eds.), Potential Theory – Surveys and Problems. Proceedings, 1987. VIII, 271 pages. 1988.

Vol. 1345: X. Gomez-Mont, J. Seade, A. Verjovski (Eds.), Holomorphic Dynamics. Proceedings, 1986. VII, 321 pages. 1988.

Vol. 1346: O. Ya. Viro (Ed.), Topology and Geometry – Rohlin Seminar. XI, 581 pages. 1988.

Vol. 1347: C. Preston, Iterates of Piecewise Monotone Mappings on an Interval. V, 166 pages. 1988.

Vol. 1348: F. Borceux (Ed.), Categorical Algebra and its Applications. Proceedings, 1987. VIII, 375 pages. 1988.

Vol. 1349: E. Novak, Deterministic and Stochastic Error Bounds in Numerical Analysis. V, 113 pages. 1988.

Vol. 1350: U. Koschorke (Ed.), Differential Topology. Proceedings, 1987. VI, 269 pages. 1988.

Vol. 1351: I. Laine, S. Rickman, T. Sorvali, (Eds.), Complex Analysis, Joensuu 1987. Proceedings. XV, 378 pages. 1988.

Vol. 1352: L.L. Avramov, K.B. Tchakerian (Eds.), Algebra – Some Current Trends. Proceedings, 1986. IX, 240 Seiten. 1988.

Vol. 1353: R.S. Palais, Ch.-l. Terng, Critical Point Theory and Submanifold Geometry. X, 272 pages. 1988.

Vol. 1354: A. Gómez, F. Guerra, M.A. Jiménez, G. López (Eds.), Approximation and Optimization. Proceedings, 1987. VI, 280 pages. 1988.

Vol. 1355: J. Bokowski, B. Sturmfels, Computational Synthetic Geometry. V, 168 pages. 1989.

Vol. 1356: H. Volkmer, Multiparameter Eigenvalue Problems and Expansion Theorems. VI, 157 pages. 1988.

Vol. 1357: S. Hildebrandt, R. Leis (Eds.), Partial Differential Equations and Calculus of Variations. VI, 423 pages. 1988.

Vol. 1358: D. Mumford, The Red Book of Varieties and Schemes. V, 309 pages. 1988.

Vol. 1359: P. Eymard, J.-P. Pier (Eds.), Harmonic Analysis. Proceedings, 1987. VIII, 287 pages. 1988.

Vol. 1360: G. Anderson, C. Greengard (Eds.), Vortex Methods. Proceedings, 1987. V, 141 pages. 1988.

Vol. 1361: T. tom Dieck (Ed.), Algebraic Topology and Transformation Groups. Proceedings, 1987. VI, 298 pages. 1988.

Vol. 1362: P. Diaconis, D. Elworthy, H. Föllmer, E. Nelson, G.C. Papanicolaou, S.R.S. Varadhan. École d'Été de Probabilités de Saint-Flour XV–XVII, 1985–87. Editor: P.L. Hennequin. V, 459 pages. 1988.

Vol. 1363: P.G. Casazza, T.J. Shura. Tsirelson's Space. VIII, 204 pages. 1988.

Vol. 1364: R.R. Phelps, Convex Functions, Monotone Operators and Differentiability. IX, 115 pages. 1989.

Vol. 1365: M. Giaquinta (Ed.), Topics in Calculus of Variations. Seminar, 1987. X, 196 pages. 1989.

Vol. 1366: N. Levitt, Grassmannians and Gauss Maps in PL-Topology. V, 203 pages. 1989.

Vol. 1367: M. Knebusch, Weakly Semialgebraic Spaces. XX, 376 pages. 1989.

Vol. 1368: R. Hübl, Traces of Differential Forms and Hochschild Homology. III, 111 pages. 1989.

Vol. 1369: B. Jiang, Ch.-K. Peng, Z. Hou (Eds.), Differential Geometry and Topology. Proceedings, 1986–87. VI, 366 pages. 1989.

Vol. 1370: G. Carlsson, R.L. Cohen, H.R. Miller, D.C. Ravenel (Eds.), Algebraic Topology. Proceedings, 1986. IX, 456 pages. 1989.

Vol. 1371: S. Glaz, Commutative Coherent Rings. XI, 347 pages. 1989.

Vol. 1372: J. Azéma, P.A. Meyer, M. Yor (Eds.), Séminaire de Probabilités XXIII. Proceedings. IV, 583 pages. 1989.

Vol. 1373: G. Benkart, J.M. Osborn (Eds.), Lie Algebras, Madison 1987. Proceedings. V, 145 pages. 1989.

Vol. 1374: R.C. Kirby, The Topology of 4-Manifolds. VI, 108 pages. 1989.

Vol. 1375: K. Kawakubo (Ed.), Transformation Groups. Proceedings, 1987. VIII, 394 pages, 1989.

Vol. 1376: J. Lindenstrauss, V.D. Milman (Eds.), Geometric Aspects of Functional Analysis. Seminar (GAFA) 1987–88. VII, 288 pages. 1989.

Vol. 1377: J.F. Pierce, Singularity Theory, Rod Theory, and Symmetry-Breaking Loads. IV, 177 pages. 1989.

Vol. 1378: R.S. Rumely, Capacity Theory on Algebraic Curves. III, 437 pages. 1989.

Vol. 1379: H. Heyer (Ed.), Probability Measures on Groups IX. Proceedings, 1988. VIII, 437 pages. 1989

Vol. 1380: H.P. Schlickewei, E. Wirsing (Eds.), Number Theory, Ulm 1987. Proceedings. V, 266 pages. 1989.

Vol. 1381: J.-O. Strömberg, A. Torchinsky. Weighted Hardy Spaces. V, 193 pages. 1989.

Vol. 1382: H. Reiter, Metaplectic Groups and Segal Algebras. XI, 128 pages. 1989.

Vol. 1383: D.V. Chudnovsky, G.V. Chudnovsky, H. Cohn, M.B. Nathanson (Eds.), Number Theory, New York 1985–88. Seminar. V, 256 pages. 1989.

Vol. 1384: J. Garcia-Cuerva (Ed.), Harmonic Analysis and Partial Differential Equations. Proceedings, 1987. VII, 213 pages. 1989.

Vol. 1385: A.M. Anile, Y. Choquet-Bruhat (Eds.), Relativistic Fluid Dynamics. Seminar, 1987. V, 308 pages. 1989.

Vol. 1386: A. Bellen, C.W. Gear, E. Russo (Eds.), Numerical Methods for Ordinary Differential Equations. Proceedings, 1987. VII, 136 pages. 1989.

Vol. 1387: M. Petković, Iterative Methods for Simultaneous Inclusion of Polynomial Zeros. X, 263 pages. 1989.

Vol. 1388: J. Shinoda, T.A. Slaman, T. Tugué (Eds.), Mathematical Logic and Applications. Proceedings, 1987. V, 223 pages. 1989.

Vol. 1000: Second Edition. H. Hopf, Differential Geometry in the Large. VII, 184 pages. 1989.

Vol. 1389: E. Ballico, C. Ciliberto (Eds.), Algebraic Curves and Projective Geometry. Proceedings, 1988. V, 288 pages. 1989.

Vol. 1390: G. Da Prato, L. Tubaro (Eds.), Stochastic Partial Differential Equations and Applications II. Proceedings, 1988. VI, 258 pages. 1989.

Vol. 1391: S. Cambanis, A. Weron (Eds.), Probability Theory on Vector Spaces IV. Proceedings, 1987. VIII, 424 pages. 1989.

Vol. 1392: R. Silhol, Real Algebraic Surfaces. X, 215 pages. 1989.

Vol. 1393: N. Bouleau, D. Feyel, F. Hirsch, G. Mokobodzki (Eds.), Séminaire de Théorie du Potentiel Paris, No. 9. Proceedings. VI, 265 pages. 1989.

Vol. 1394: T.L. Gill, W.W. Zachary (Eds.), Nonlinear Semigroups, Partial Differential Equations and Attractors. Proceedings, 1987. IX, 233 pages. 1989.

Vol. 1395: K. Alladi (Ed.), Number Theory, Madras 1987. Proceedings. VII, 234 pages. 1989.

Vol. 1396: L. Accardi, W. von Waldenfels (Eds.), Quantum Probability and Applications IV. Proceedings, 1987. VI, 355 pages. 1989.

Vol. 1397: P.R. Turner (Ed.), Numerical Analysis and Parallel Processing. Seminar, 1987. VI, 264 pages. 1989.

Vol. 1398: A.C. Kim, B.H. Neumann (Eds.), Groups – Korea 1988. Proceedings. V, 189 pages. 1989.

Vol. 1399: W.-P. Barth, H. Lange (Eds.), Arithmetic of Complex Manifolds. Proceedings, 1988. V, 171 pages. 1989.

Vol. 1400: U. Jannsen. Mixed Motives and Algebraic K-Theory. XIII, 246 pages. 1990.

Vol. 1401: J. Steprāns, S. Watson (Eds.), Set Theory and its Applications. Proceedings, 1987. V, 227 pages. 1989.

Vol. 1402: C. Carasso, P. Charrier, B. Hanouzet, J.-L. Joly (Eds.), Nonlinear Hyperbolic Problems. Proceedings, 1988. V, 249 pages. 1989.

Vol. 1403: B. Simeone (Ed.), Combinatorial Optimization. Seminar, 1986. V, 314 pages. 1989.

Vol. 1404: M.-P. Malliavin (Ed.), Séminaire d'Algèbre Paul Dubreil et Marie-Paul Malliavin. Proceedings, 1987 – 1988. IV, 410 pages. 1989.

Vol. 1405: S. Dolecki (Ed.), Optimization. Proceedings, 1988. V, 223 pages. 1989.

Vol. 1406: L. Jacobsen (Ed.), Analytic Theory of Continued Fractions III. Proceedings, 1988. VI, 142 pages. 1989.

Vol. 1407: W. Pohlers, Proof Theory. VI, 213 pages. 1989.

Vol. 1408: W. Lück, Transformation Groups and Algebraic K-Theory. XII, 443 pages. 1989.

Vol. 1409: E. Hairer, Ch. Lubich, M. Roche. The Numerical Solution of Differential-Algebraic Systems by Runge-Kutta Methods. VII, 139 pages. 1989.

Vol. 1410: F.J. Carreras, O. Gil-Medrano, A.M. Naveira (Eds.), Differential Geometry. Proceedings, 1988. V, 308 pages. 1989.

Vol. 1411: B. Jiang (Ed.), Topological Fixed Point Theory and Applications Proceedings, 1988. VI, 203 pages. 1989.

Vol. 1412: V.V. Kalashnikov, V.M. Zolotarev (Eds.), Stability Problems fo Stochastic Models. Proceedings, 1987. X, 380 pages. 1989.

Vol. 1413: S. Wright, Uniqueness of the Injective III_1 Factor. III, 108 pages 1989.

Vol. 1414: E. Ramírez de Arellano (Ed.), Algebraic Geometry and Comple Analysis. Proceedings, 1987. VI, 180 pages. 1989.

Vol. 1415: M. Langevin, M. Waldschmidt (Eds.), Cinquante Ans de Poly nômes. Fifty Years of Polynomials. Proceedings, 1988. IX, 235 pages 1990.

Vol. 1416: C. Albert (Ed.), Géométrie Symplectique et Mécanique. V, 28 pages. 1990.

Vol. 1417: A.J. Sommese, A. Biancofiore, E.L. Livorni (Eds.). Algebrai Geometry. Proceedings, 1988. V, 320 pages. 1990.

Vol. 1418: M. Mimura, Homotopy Theory and Related Topics. Proceedings 1988. V, 241 pages. 1990.

Vol. 1419: P.S. Bullen, P.Y. Lee, J.L. Mawhin, P. Muldowney, W.F. Pfeffe (Eds.), New Integrals. Proceedings, 1988. V, 202 pages. 1990.

Vol. 1420: M. Galbiati, A. Tognoli (Eds.), Real Analytic Geometry. Procee ings, 1988. IV, 366 pages. 1990.

Vol. 1421: H.A. Biagioni, A Nonlinear Theory of Generalized Functions. X 214 pages. 1990.

Vol. 1422: V. Villani (Ed.), Complex Geometry and Analysis. Proceedings 1988. V, 109 pages. 1990.

Vol. 1423: S.O. Kochman, Stable Homotopy Groups of Spheres: A Con puter-Assisted Approach. VIII, 330 pages. 1990.

Vol. 1424: F.E. Burstall, J.H. Rawnsley, Twistor Theory for Riemanni Symmetric Spaces. III, 112 pages. 1990.